LONDON MATHEMATICAL SOCIETY LECTURE NOTE SERIES

Managing Editor: Professor J.W.S. Cassels,
Department of Pure Mathematics and Mathematical Statistics,
16 Mill Lane, Cambridge CB2 1SB.

1. General cohomology theory and K-theory, P.HILTON
4. Algebraic topology, J.F.ADAMS
5. Commutative algebra, J.T.KNIGHT
8. Integration and harmonic analysis on compact groups, R.E.EDWARDS
9. Elliptic functions and elliptic curves, P.DU VAL
10. Numerical ranges II, F.F.BONSALL & J.DUNCAN
11. New developments in topology, G.SEGAL (ed.)
12. Symposium on complex analysis, Canterbury, 1973, J.CLUNIE & W.K.HAYMAN (eds.)
13. Combinatorics: Proceedings of the British Combinatorial Conference 1973, T.P.McDONOUGH & V.C.MAVRON (eds.)
15. An introduction to topological groups, P.J.HIGGINS
16. Topics in finite groups, T.M.GAGEN
17. Differential germs and catastrophes, Th.BROCKER & L.LANDER
18. A geometric approach to homology theory, S.BUONCRISTIANO, C.P. ROURKE & B.J.SANDERSON
20. Sheaf theory, B.R.TENNISON
21. Automatic continuity of linear operators, A.M.SINCLAIR
23. Parallelisms of complete designs, P.J.CAMERON
24. The topology of Stiefel manifolds, I.M.JAMES
25. Lie groups and compact groups, J.F.PRICE
26. Transformation groups: Proceedings of the conference in the University of Newcastle-upon-Tyne, August 1976, C.KOSNIOWSKI
27. Skew field constructions, P.M.COHN
28. Brownian motion, Hardy spaces and bounded mean oscillations, K.E.PETERSEN
29. Pontryagin duality and the structure of locally compact Abelian groups, S.A.MORRIS
30. Interaction models, N.L.BIGGS
31. Continuous crossed products and type III von Neumann algebras, A.VAN DAELE
32. Uniform algebras and Jensen measures, T.W.GAMELIN
33. Permutation groups and combinatorial structures, N.L.BIGGS & A.T.WHITE
34. Representation theory of Lie groups, M.F. ATIYAH et al.
35. Trace ideals and their applications, B.SIMON
36. Homological group theory, C.T.C.WALL (ed.)
37. Partially ordered rings and semi-algebraic geometry, G.W.BRUMFIEL
38. Surveys in combinatorics, B.BOLLOBAS (ed.)
39. Affine sets and affine groups, D.G.NORTHCOTT
40. Introduction to Hp spaces, P.J.KOOSIS
41. Theory and applications of Hopf bifurcation, B.D.HASSARD, N.D.KAZARINOFF & Y-H.WAN
42. Topics in the theory of group presentations, D.L.JOHNSON
43. Graphs, codes and designs, P.J.CAMERON & J.H.VAN LINT
44. Z/2-homotopy theory, M.C.CRABB
45. Recursion theory: its generalisations and applications, F.R.DRAKE & S.S.WAINER (eds.)
46. p-adic analysis: a short course on recent work, N.KOBLITZ
47. Coding the Universe, A.BELLER, R.JENSEN & P.WELCH
48. Low-dimensional topology, R.BROWN & T.L.THICKSTUN (eds.)

49. Finite geometries and designs, P.CAMERON, J.W.P.HIRSCHFELD & D.R.HUGHES (eds.)
50. Commutator calculus and groups of homotopy classes, H.J.BAUES
51. Synthetic differential geometry, A.KOCK
52. Combinatorics, H.N.V.TEMPERLEY (ed.)
53. Singularity theory, V.I.ARNOLD
54. Markov processes and related problems of analysis, E.B.DYNKIN
55. Ordered permutation groups, A.M.W.GLASS
56. Journées arithmétiques 1980, J.V.ARMITAGE (ed.)
57. Techniques of geometric topology, R.A.FENN
58. Singularities of smooth functions and maps, J.MARTINET
59. Applicable differential geometry, M.CRAMPIN & F.A.E.PIRANI
60. Integrable systems, S.P.NOVIKOV et al.
61. The core model, A.DODD
62. Economics for mathematicians, J.W.S.CASSELS
63. Continuous semigroups in Banach algebras, A.M.SINCLAIR
64. Basic concepts of enriched category theory, G.M.KELLY
65. Several complex variables and complex manifolds I, M.J.FIELD
66. Several complex variables and complex manifolds II, M.J.FIELD
67. Classification problems in ergodic theory, W.PARRY & S.TUNCEL
68. Complex algebraic surfaces, A.BEAUVILLE
69. Representation theory, I.M.GELFAND et al.
70. Stochastic differential equations on manifolds, K.D.ELWORTHY
71. Groups - St Andrews 1981, C.M.CAMPBELL & E.F.ROBERTSON (eds.)
72. Commutative algebra: Durham 1981, R.Y.SHARP (ed.)
73. Riemann surfaces: a view towards several complex variables, A.T.HUCKLEBERRY
74. Symmetric designs: an algebraic approach, E.S.LANDER
75. New geometric splittings of classical knots (algebraic knots), L.SIEBENMANN & F.BONAHON
76. Linear differential operators, H.O.CORDES
77. Isolated singular points on complete intersections, E.J.N.LOOIJENGA
78. A primer on Riemann surfaces, A.F.BEARDON
79. Probability, statistics and analysis, J.F.C.KINGMAN & G.E.H.REUTER (eds.)
80. Introduction to the representation theory of compact and locally compact groups, A.ROBERT
81. Skew fields, P.K.DRAXL
82. Surveys in combinatorics: Invited papers for the ninth British Combinatorial Conference 1983, E.K.LLOYD (ed.)
83. Homogeneous structures on Riemannian manifolds, F.TRICERRI & L.VANHECKE
84. Finite group algebras and their modules, P.LANDROCK
85. Solitons, P.G.DRAZIN
86. Topological topics, I.M.JAMES (ed.)
87. Surveys in set theory, A.R.D.MATHIAS (ed.)
88. FPF ring theory, C.FAITH & S.PAGE
89. An F-space sampler, N.J.KALTON, N.T.PECK & J.W.ROBERTS
90. Polytopes and symmetry, S.A.ROBERTSON

London Mathematical Society Lecture Note Series: 90

Polytopes and Symmetry

STEWART A. ROBERTSON

Professor of Pure Mathematics,
University of Southampton

CAMBRIDGE UNIVERSITY PRESS

Cambridge

London New York New Rochelle

Melbourne Sydney

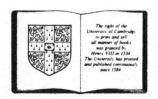

Published by the Press Syndicate of the University of Cambridge
The Pitt Building, Trumpington Street, Cambridge CB2 1RP
32 East 57th Street, New York, NY 10022, USA
296 Beaconsfield Parade, Middle Park, Melbourne 3206, Australia

First published 1984

Library of Congress catalogue card number: 83-15171

British Library Cataloguing in Publication Data

Robertson, Stewart A.
Polytopes and symmetry - (London Mathematical Society lecture note series, ISSN 0076-0552; 90)
1. Polytopes
I. Title II. Series
512'.33 QA691

ISBN 0 521 27739 6

Transferred to digital printing 2001

Contents

Preface

These notes are intended to give a fairly systematic exposition of an approach to the symmetry classification of convex polytopes that casts some fresh light on classical ideas and generates a number of new theorems. The theory is far from complete, and there is a range of attractive unsolved problems. I hope that the level of sophistication will be found suitable for anyone with a knowledge of contemporary pure mathematics at first degree level.

The work has developed sporadically over almost twenty years. In the early stages, I worked in collaboration with Sheila Carter and Hugh Morton at Liverpool. At that time we concentrated on highly symmetric polytopes, but many of the key ideas apply equally well to the study of polytopes in general. We were greatly helped in these days by the generous advice and encouragement of C.T.C. Wall. Morton's influence is particularly strong in Chapter 4, which is based on previously unpublished joint work, while Chapter 6 owes much to Sheila Carter's patient investigation of symmetric polyhedra. Since coming to Southampton in 1970, I have received the benefit of A.W. Deicke's acute geometrical insight. Deicke's contributions to my understanding of cuboids and polyhedra generally have been considerable. Other colleagues, notably David Chillingworth, Gareth Jones and David Singerman, have helped me by producing good examples and neat lines of argument in response to my questions. In recent years, D.G. Kendall has been developing a theory of shapes for finite sets of points in Euclidean space that has something in common with my approach to polytopes, although Kendall is concerned with a rather different class of problems. His concern is with statistical questions, and convexity does not hold the central position that I have given it here. Kendall's work makes considerable use of metrical ideas, which I have used only to generate topologies. Nevertheless, there is a common spirit in the two theories, and the reader who sees anything of interest in this account will certainly derive great benefit from a study of Kendall's work. See Kendall (1983) for a recent version and references to previous and forthcoming articles.

Another substantial piece of work that is related to my approach to polytopes is Schwarzenberger's study of crystallography in n-dimensional space (Schwarzenberger (1980)).

Schwarzenberger gives a systematic treatment of the concepts needed to provide a coherent framework for the classification of 'crystal' structures in Euclidean n-space E^n. The geometrical structures involved are lattices in E^n rather than polytopes, and the groups are infinite discrete subgroups of the Euclidean group rather than finite subgroups of the orthogonal group. The situation in crystallography is more difficult to handle than in the theory of polytopes, since the symmetry groups themselves admit continuous deformations.

I thank I.M. James for his patient encouragement, June Kerry for her efficient work in preparing the typescript, and D.G. Kendall for his friendly attitude in giving me access to some of his unpublished work.

Southampton
June, 1983.

S.A. Robertson

0. SYNOPSIS

The following is a brief outline of some of the main ideas and results that are discussed in detail in the main body of the text. There is no direct relation between the organisation of the paragraphs here and the arrangement of the various chapters.

1. Euclidean space

The polytopes discussed in this book are convex polytopes in Euclidean n-space E^n , where n is any positive integer. So that all polytopes can be treated simultaneously as members of a single family, it is convenient to regard E^n as a fixed linear subspace of E^{n+1} . Consequently, we employ the union E of all the spaces E^n as a receptacle for all polytopes, referring to E as Euclidean space.

2. Polytopes

Our objects of study are the convex hulls $P = \text{conv } A$ of the finite subsets A of E . The dimension $\dim P$ of a polytope $P = \text{conv } A$ is the dimension of its affine hull $\text{aff } P = \text{aff } A$, and P is called an n-polytope if $\dim P = n$. Thus the empty set $\emptyset$ is the unique (-1)-polytope, while 0-polytopes are singleton subsets of E , 1-polytopes are closed bounded straight line segments in E , 2-polytopes are (convex, plane) polygons, and 3-polytopes are (convex) polyhedra.

3. Symmetry

The broad aim of the theory is the classification of polytopes according to their geometrical symmetry. The first step in the creation of such a theory is the choice of a basic equivalence relation $\simeq$, say, on the family $\mathcal{P}$ of all polytopes. We must choose $\simeq$ in such a way that it captures what is meant by the statement that two polytopes 'have the same geometrical symmetry'. If P is similar to Q , written $P \sim Q$, that is to say if P has exactly the same shape as Q , differing only in size and position within E , then surely $P \simeq Q$. For instance, if P and Q are plane rectangles in which the ratios of the longer to the shorter sides are λ_P and λ_Q respectively, then $P \sim Q$ iff $\lambda_P = \lambda_Q$. However, we expect that $P \simeq Q$ iff λ_P and λ_Q are both greater than 1 or both equal to 1.

4. Combinatorial equivalence

The combinatorial theory of polytopes takes no account of metrical structure, but regards the face-lattice $F(P)$ of any polytope P as its principal structural feature. The appropriate equivalence relation in this context is therefore combinatorial equivalence $\overset{\sim}{\sim}$, where $P \overset{\sim}{\sim} Q$ means that $F(P)$ is isomorphic to $F(Q)$. The lattice $F(P)$ is made up of the various <u>i-faces</u> of P , $i = -1,0,1,\ldots$, $n = \dim P$. Its minimal element is $\emptyset$ and its maximal element is P itself. The partial order in $F(P)$ is inclusion. Each i-face is itself an i-polytope, and for any polytopes P , P' we write $P' \lhd P$ to mean that P' is a face of P . Among the obvious combinatorial invariants of a polytope P are the numbers $f_i(P)$ of i-faces, $i = 0,\ldots, n-1$, and the group $\mathrm{Aut}\, P$ of automorphisms of $F(P)$ (up to group isomorphism). Another useful invariant is the number $\mu(P)$ of pairs (v,T) , where v is a 0-face or <u>vertex</u> of P , T is an (n-1)-face or <u>facet</u> of P , and $v < T$. I call $\mu(P)$ the <u>multiplicity</u> of P .

5. Symmetry equivalence

I take the view that the geometrical symmetry properties of a polytope P are embodied in the action α_P of the <u>symmetry group</u> $\Gamma(P)$ of P on the face lattice $F(P)$. The group $\Gamma(P)$ is the group of all rigid transformations of aff P that leave P setwise fixed. Thus $\Gamma(P)$ is a finite group that embeds naturally in the finite group $\mathrm{Aut}\, P$. I say therefore that P is <u>symmetry equivalent</u> to Q , and write $P \simeq Q$, iff the action α_P is equivalent to α_Q . This means that there is a conjugacy γ from $\Gamma(P)$ to $\Gamma(Q)$ in the Euclidean group Iso and an isomorphism λ from $F(P)$ to $F(Q)$ such that, for all $g \in \Gamma(P)$ and all $K \in F(P)$, $\lambda(g.K) = \gamma(g).\lambda(K)$. It follows at once that, for all polytopes P and Q ,

$$P \sim Q \Rightarrow P \simeq Q \Rightarrow P \overset{\sim}{\sim} Q ,$$

none of the implications being reversible, as illustrated in the following sequence of quadrilaterals.

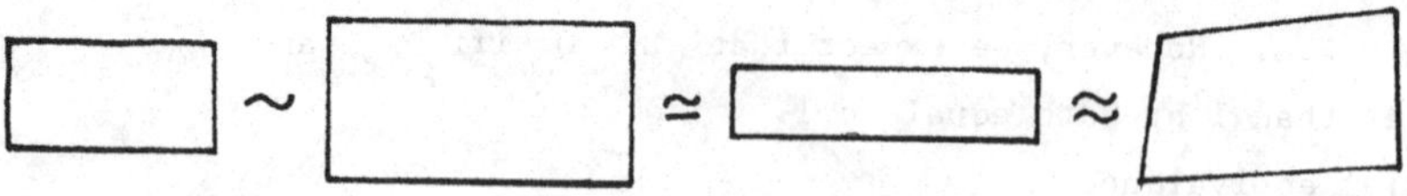

6. Topology

The Hausdorff metric can be exploited to endow the family of all polytopes with a topology in which, leaving the empty set and the singletons aside, the family $\mathcal{P}^+$ of all n-polytopes with $n \geq 1$ is an arcwise connected metric space. The quotient space $\mathfrak{S}^+ = \mathcal{P}^+/\sim$ is given the quotient topology, and the similarity classes of the empty set and of 0-polytopes are adjoined to $\mathfrak{S}^+$ as two singleton components $\oslash$ and $\odot$, to give the space $\mathfrak{S} = \oslash \cup \odot \cup \mathfrak{S}^+$ of all similarity classes of polytopes. The relations $\approx$ and $\overset{\sim}{\sim}$ may be carried down to $\mathfrak{S}$.

7. The symmetry type stratification

Each of the symmetry types into which symmetry equivalence partitions $\mathfrak{S}$ is a topological manifold whose frontier in $\mathfrak{S}$ is a union of finitely many symmetry types of lower dimension. To this extent, $\approx$ determines a 'topological stratification' of $\mathfrak{S}$. For example, there is a symmetry type of dimension 1 consisting of the similarity classes of all nonsquare rectangles. Its frontier in $\mathfrak{S}$ consists of the 0-dimensional symmetry types of all squares and of all 1-polytopes.

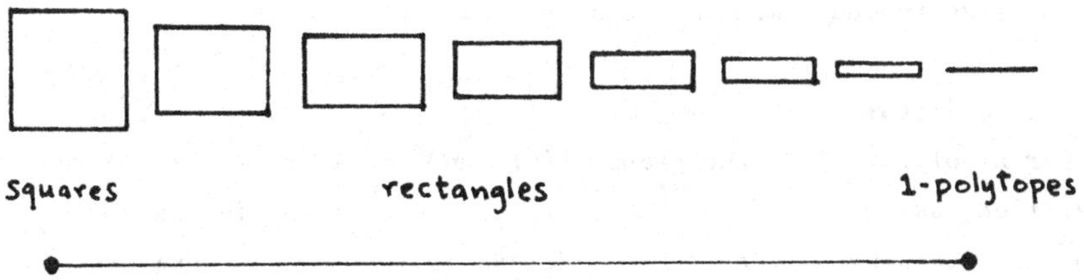

8. Deficiency

For each polytope P we try to find relations between the action α_P and the topological invariants of the symmetry type $\mathcal{S}(P)$ to which (the similarity class of) P belongs. Little progress has been made on this problem. However, there is plenty of evidence to suggest that the dimension $\dim \mathcal{S}(P)$ of $\mathcal{S}(P)$ is a good measure of how 'unsymmetrical' P is. Thus the 'most highly symmetrical' polytopes are those for which $\dim \mathcal{S}(P) = 0$. Such polytopes are said to be <u>perfect</u> (or of <u>perfect form</u>, or <u>perfectly symmetrical</u>). I also refer to $\dim \mathcal{S}(P)$ as the <u>deficiency</u> of P and denote this number by def P. For instance, a square has deficiency 0 and is therefore an example of a perfect polygon. A rectangle has deficiency 1.

9. Regularity and perfection

Every polytope that is regular in the classical sense is also perfect. For example, the set of all perfect polygons coincides with the set of all regular polygons. Likewise all five Platonic solids are perfect. There are, however, four other similarity classes of perfect polyhedra, represented by the cuboctahedron, the icosidodecahedron and their polars the rhombic dodecahedron of the first kind and the rhombic triacontahedron. These nine perfect solids are, up to similarity, the only polyhedra whose symmetry group acts transitively on edges. There is a countable infinity of similarity classes of perfect 4-polytopes, and indeed of perfect 2m-polytopes, for every $m \geq 1$ (see §13 below). Whether this is so in odd dimensions ≥ 5 has not been settled.

10. Deicke's conjecture

For a polygon P, the group $\Gamma(P)$ and the lattice $F(P)$ are easily specified, as is the action α_P. In each one of the few cases that can occur, it is easy to calculate def P in terms of the numbers of vertex- and edge-orbits of α_P. For polytopes of higher dimension, however, the problem is more complicated. For polyhedra, the following conjecture has been proposed by A.W. Deicke.

DEICKE'S CONJECTURE Let P be any polyhedron. Suppose that the action of $\Gamma(P)$ on the set of edges of P has e orbits. Then def P = e − 1.

For example, Deicke's conjecture implies that P is perfect iff $e = 1$, which we have shown to be the case by ad hoc methods. Again, if $\Gamma(P)$ is trivial, then $e = f_o(P)$ is just the number of edges of P, and a combination of Euler's formula and the formula given in §11 below yields $\operatorname{def} P = f_o(P) - 1 = e - 1$ in this case.

11. Combinatorial types

If P is an n-polytope in E^n, then the set $\mathbb{C}_n P$ of all polytopes Q in E^n such that $Q \approx P$ is a subset of the space $\mathcal{P}$. A straightforward application of the implicit function theorem shows that $\mathbb{C}_n P$ has the structure of a differentiable manifold of dimension $n(r+s) - m$, where $r = f_o(P)$ is the number of vertices of P, $s = f_{n-1}(P)$ is the number of facets of P, and $m = \mu(P)$ is the multiplicity of P (see §4). In case P is a cube, $r = 8$, $s = 6$, $m = 24$ and $\dim \mathbb{C}_3 P = 18$. The set $\mathcal{N}_P$ of polytopes in $\mathbb{C}_n P$ whose centroid is 0 and whose 'radius' is 1 is a compact submanifold of $\mathbb{C}_n P$ of codimension $n + 1$. The contribution of $\mathbb{C}_n P$ to the space $\mathfrak{S}$ can then be identified with the quotient space $\mathcal{N}_P/O(n)$ of $\mathcal{N}_P$ by the orthogonal group $O(n)$ in E^n, and the orbit types of this action are identified with the symmetry types in this part of $\mathfrak{S}$. In particular, the principal orbit type corresponds to the symmetry type of polytopes $Q \approx P$ for which $\Gamma(Q)$ is trivial.

For example, if P is a cube in E^3, then there are 22 symmetry types of 'cuboids' $Q \approx P$ ranging in dimension $\operatorname{def} Q$ from 0 (P itself) to $11 = \dim[P]_3 - 4 - \dim O(3)$. There are no cuboids Q with $\operatorname{def} Q = 8, 9$ or 10.

12. Duality and polarity

If P and Q are polytopes, then P is said to be dual to Q, written $P \wr\wr Q$, iff there is an anti-isomorphism from $F(P)$ to $F(Q)$. For example, a cube is dual to an octahedron, and a tetrahedron is self-dual. For any n-polytope P in E^n, let c be the centroid of P and S_r the sphere of centre c and radius $r > 0$ in E^n. There is a unique n-polytope P_r^* whose facets lie in the polars with respect to S_r of the vertices of P. Then $P_r^* \wr\wr P$ for all $r > 0$. In fact, $P_r^* \sim P_s^*$ for all $r,s > 0$, and $(P_r^*)_s^* \approx P$. Note that $\wr\wr$

is symmetric but is neither reflexive nor transitive. We call P_r^* the r-polar of P . It follows from these remarks that there is a polarity map $\pi : \mathfrak{S} \to \mathfrak{S}$ sending the similarity class of P to that of its r-polar P_r^* . However, π is not continuous (see §2.6). To get round this unpleasant fact, we consider the space $\mathfrak{S}(n)$ of all similarity classes of m-polytopes with $m \leqslant n$, and form the quotient space $\mathfrak{S}^{\dagger}(n) = \mathfrak{S}(n)/\mathfrak{S}(n-1)$, collapsing $\mathfrak{S}(n-1)$ to a single point. Then symmetry equivalence is well-defined on $\mathfrak{S}^{\dagger}(n)$ and π induces a polarity map $\tilde{\pi}_n : \mathfrak{S}^{\dagger}(n) \to \mathfrak{S}^{\dagger}(n)$ which respects $\approx$. Let $\mathcal{G}_n$ denote the group of all homeomorphisms of $\mathfrak{S}^{\dagger}(n)$ to itself preserving symmetry equivalence, modulo those that map each symmetry type to itself. Then $\mathcal{G}_n$ is a discrete group to which $\tilde{\pi}_n$ contributes an element of order 2 for each $n \geqslant 2$. It seems unlikely to me that $\mathcal{G}_n$ contains any other nontrivial elements, and so I propose the following conjecture.

POLARITY CONJECTURE For all $n \geqslant 2$, $\mathcal{G}_n$ is cyclic of order 2 .

13. Products and sums

The concept of rectangular product for polytopes is long-established, having been introduced by Schoute at the end of the last century. A rectangle, for example, is the rectangular product of two 1-polytopes. We denote this binary operation by $\square$, and show that every polytope has a decomposition $P_1 \square \ldots \square P_r$ where each polytope P_i is $\square$ -indecomposable. This decomposition is unique up to the order of the factors, and so there is a notion of $\square$ -primeness for polytopes. It is not difficult to express $\Gamma(P)$ in terms of $\Gamma(P_i)$, and def(P) in terms of def(P_i) , $i = 1,\ldots,r$. In particular, if P is perfect, then so is $\square^k P = P \square \ldots \square P$ (k factors) for any $k \geqslant 2$. Thus the hypercube $\square_k = \square^k I$, where I is any 1-polytope, is perfect for all $k \geqslant 2$. There is a dual or polar concept of rectangular sum $\diamond$. An octahedron may be regarded as the rectangular sum of three 1-polytopes. For any polytopes P and Q , $P \wr\wr P'$ and $Q \wr\wr Q'$ imply that $P \square Q \wr\wr P' \diamond Q'$, and this proposition, in a somewhat refined form, applies to polars: the cross-polytope $\diamond_k = \diamond^k I$ may be regarded as the polar of $\square_k$. Although the operators $\square$ and $\diamond$ apply to congruence classes of polytopes rather than to similarity classes, they are useful in the study of $\mathfrak{S}$, especially in the search for perfect polytopes. Since every regular polygon P is perfect, for instance, so also are $\Box^k P$ and $\Diamond^k P$. Hence we know at once that there is a countable infinity of similarity classes of 2k-polytopes, for each $k \geq 1$.

1. THE SPACE OF POLYTOPES

Polytopes are closely related to two other families of subsets of Euclidean space, namely finite subsets and affine planes. We explore this relationship and construct a natural topology for each of the first two families. The basic equivalence relation of similarity and its interpretation in terms of transformation groups are also discussed.

1. Euclidean space

For any positive integer n, Euclidean n-space will be denoted by E^n. It is convenient to represent E^n as a fixed subspace of E^{n+1}, and we do this by identifying $x \in E^n$ with $Y \in E^{n+1}$, where $y_i = x_i$ for $1 \leq i \leq n$ and $y_{n+1} = 0$. Likewise, we consider the space E of all infinite sequences $x = (x_i) = (x_1, \ldots, x_n, \ldots)$ of real numbers x_i in which $x_i = 0$ for all but finitely many values of i, and we embed E^n in E by identifying $x \in E^n$ with $z \in E$, where $z_i = x_i$ for $1 \leq i \leq n$ and $z_i = 0$ for $i > n$. In this way we obtain a sequence of inclusions

$$E^1 \subset E^2 \subset \ldots \quad \subset E^n \subset E^{n+1} \subset \ldots$$

and we can write $E = \bigcup_{n=1}^{\infty} E^n = \lim\limits_{\rightarrow} E^n$.

The space E has a real linear structure given by

$$\lambda(x_i) + \mu(y_i) = (z_i),$$

where $z_i = \lambda x_i + \mu y_i$, $i = 1, \ldots, n, \ldots$, for $\lambda, \mu \in R$, $x = (x_i)$, $y = (y_i) \in E$. The metrical structure of E is determined by the familiar Euclidean inner product $<\,,\,>$, where $<x,y> = \Sigma_{i=1}^{\infty} x_i y_i$ for all $x,y \in E$. The associated norm $\|\ \|$ and Euclidean metric d are then given by $\|x\| = \sqrt{<x,x>}$ and $d(x,y) = \|x - y\|$ respectively.

The resulting metric space (E,d), which we denote simply by E, is called Euclidean space. The fact that E is not complete will cause us no trouble, since we are interested only in subsets of E that lie in some E^n. Our only reason for introducing E is to be able to handle all the spaces E^n at once.

2. Affine hulls of finite sets

Let $\mathcal{F}$ denote the family of all finite subsets of E. We denote the number of elements in $A \in \mathcal{F}$ by $\#A$ (and extend this notation to finite sets in general in Chapter 2). Thus $\#\emptyset = 0$ and $\#A = 1$ iff A is a singleton $\{a\}$, $a \in E$.

If $A \in \mathcal{F}$, then any linear combination $\Sigma_{a\in A}\, t_a a$ for which $\Sigma_{a\in A} t_a = 1$ is called an <u>affine combination</u> of A. The set of all affine combinations of A is called the <u>affine hull</u> aff A of A, and any such subset $X = \text{aff}A$ of E is called an <u>affine plane</u> generated by A. More generally, for any subset S of E, affS denotes the union of the affine planes affA, where $A \in \mathcal{F}$ and $A \subset S$.

Let $\mathcal{A}$ denote the set of all affine planes of the form affA, $A \in \mathcal{F}$. Then $\text{aff}: \mathcal{F} \to \mathcal{A}$ is a surjective map. For example, $\text{aff}A = A$ iff $A = \emptyset$ or A is a singleton $\{a\}$. Of course it may happen that $\text{aff}A = \text{aff}B$ for distinct sets $A, B \in \mathcal{F}$. If $\text{aff}A = X$ and for every proper subset S of A, aff S is a proper subset of X, then A is said to be <u>affinely independent</u> and to be an <u>affine basis</u> for X.

The set $\mathcal{L}$ of all finite-dimensional linear subspaces of E is a subset of $\mathcal{A}$ with inclusion $\iota: \mathcal{L} \to \mathcal{A}$, say. For convenience, we give $\emptyset$ 'honorary membership' of $\mathcal{L}$, assigning to it the dimension -1. There is then a parallel projection $p: \mathcal{A} \to \mathcal{L}$ given by $p(X) = L$, where $L = \{x-y : x, y \in X\}$. Trivially, $p\circ\iota = 1_{\mathcal{L}}$. We say that $X, Y \in \mathcal{A}$ are <u>parallel,</u> written $X \| Y$, iff $P(X) \subset P(Y)$ or $P(Y) \subset P(X)$.

The dimension dim L of any $L \in \mathcal{L}$ is well-defined, and we extend this notion to the elements of $\mathcal{A}$ by putting $\dim X = \dim P(X)$ for each $X \in \mathcal{A}$. We say that $X \in \mathcal{A}$ is an <u>affine k-plane</u> iff $\dim X = k$.

Exercise: Let $A \in \mathcal{F}$ and $X = \text{aff}A$. Then $\#A \geq 1 + \dim X$ with equality iff A is affinely independent.

Exercise: Let $A \in \mathcal{F}$. Then $A \subset E^n$ for some n. Hence $\text{aff}A \subset E^n$. Thus if $\mathcal{F}_n$ and $\mathcal{A}_n$ denote the sets of all finite subsets and of all affine planes in E^n, respectively, then $\mathcal{F} = \bigcup_{n\geq 1} \mathcal{F}_n$ and $\mathcal{A} = \bigcup_{n\geq 1} \mathcal{A}_n$.

3. Lattice structure

Both $\mathcal{F}$ and $\mathcal{A}$ inherit the partial order by inclusion from the set 2^E of all subsets of E, and both are lattices with respect to this partial order. In $\mathcal{F}$, the least upper bound and greatest lower bound of $A,B \in \mathcal{F}$ are $A \cup B$ and $A \cap B$ respectively. In $\mathcal{A}$, we denote the least upper bound of X and Y by $X \uplus Y$, while the greatest lower bound of $X, Y \in \mathcal{A}$ is just $X \cap Y$. The fact is that $X \cup Y = X \uplus Y$ iff $X \subset Y$ or $Y \supset X$, since $X \uplus Y$ is the intersection of all affine planes that contain $X \cup Y$.

Exercise: Let $A,B \in \mathcal{F}$ and $X = \text{aff}A$, $Y = \text{aff}B$. Then $A \subset B \Rightarrow X \subset Y$, $\text{aff}(A \cup B) = X \uplus Y$, and $\text{aff}(A \cap B) \subset X \cap Y$. Find A, B such that the last inclusion is proper.

Exercise: For all $L,M \in \mathcal{L}$, $L \uplus M = L + M$. For all $X,Y \in \mathcal{A}$, $p(X \uplus Y) = p(X) + p(Y)$ and $p(X \cap Y) \subset p(X) \cap p(Y)$.

4. Polytopes

Let $A \in \mathcal{F}$. Any affine combination $\Sigma_{a \in A} t_a a$ such that, for all $a \in A$, $t_a \geq 0$ is called a <u>convex combination</u> of A, and the set of all convex combinations of A is called the <u>convex hull</u> $\text{conv}\, A$ of A. Any subset $P = \text{conv}\, A$, where $A \in \mathcal{F}$, is called a <u>convex polytope</u> or simply a <u>polytope</u>. We denote the set of all polytopes by $\mathcal{P}$. It follows at once from the definitions that for all $A \in \mathcal{F}$,

$$A \subset \text{conv}\, A \subset \text{aff}A = \text{aff}(\text{conv}A) .$$

If $P = \text{conv}A \in \mathcal{P}$, then P is said to be of <u>dimension</u> $\dim P = \dim(\text{aff}A)$, and we refer to P as an n-<u>polytope</u> iff $n = \dim P$.

The empty set $\emptyset$ is the unique (-1)-polytope. The 0-polytopes are the singleton subsets of E, and 1-polytopes are closed bounded straight line-segments. A 2-polytope is called a <u>polygon</u> and a 3-polytope is called a <u>polyhedron</u>.

Notice that if $A, B \in \mathcal{F}$ and $A \subset B$, then $\text{conv}\, A \subset \text{conv}\, B$. A more subtle fact is that if $\text{conv}\, A = \text{conv}\, B = P$, then $P = \text{conv}(A \cap B)$. It follows that for each polytope P there is a unique set $V \in \mathcal{F}$ such that $P = \text{conv}\, V$ and for every proper subset W of V, $\text{conv}\, W$ is a proper subset of P. Thus V is the 'smallest' subset of E whose convex hull is P. The elements of V are called the <u>vertices</u>

of P, and V is called the <u>vertex set</u> vert P of P. The relationship between vert and conv is shown in the commutative diagram

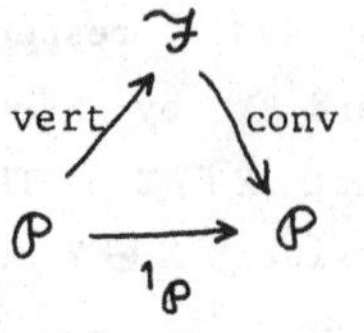

Thus vert is injective and conv is surjective.

It follows from the first Exercise of §2 that an n-polytope P has at least n + 1 vertices, and has exactly n + 1 vertices iff vert P is affinely independent. Another way of putting this is to say that an n-polytope P has exactly n + 1 vertices iff these vertices are 'in general position' in E. Such an n-polytope is called an n-<u>simplex</u>. Of course for $-1 \leqslant n \leqslant 1$, every n-polytope is an n-simplex. A 2-simplex is called a <u>triangle</u> and a 3-simplex is called a <u>tetrahedron</u>, for reasons that either are familiar already or will become clear shortly.

With the second Exercise of § 2 in mind, we note that for each $P \in \mathcal{P}$ there is a positive integer m such that $P \subset E^m$. Trivially, $m \geqslant \dim P$. We denote the set of all polytopes in E^m by $\mathcal{P}_m$ and note that $\mathcal{P} = \bigcup_{n\geq 1}\mathcal{P}_m$.

5. The Hausdorff metric

There is a well-known procedure, due to Hausdorff [Kelley, 1942; Grünbaum, 1967], that can be used here to topologise both $\mathcal{F}$ and $\mathcal{P}$ in a natural way. We describe this first in the general context of metric spaces.

Let (M,δ) be a metric space, and let $\mathcal{C}(M)$ denote the set of all nonempty compact subsets of M. The <u>Hausdorff metric</u> h on $\mathcal{C}(M)$ is defined as follows. Let $S, T \in \mathcal{C}(M)$. Then

$$h(S,T) = \max\{\lambda,\mu\},$$

where $\lambda = \max_{x\in S} \min_{y\in T} \delta(x,y)$ and $\mu = \max_{y\in T} \min_{x\in S} \delta(x,y)$. As an illustration, let $M = E^2$, $\delta = d|E^2$ and let S and T be closed circular discs with centres $(-a,0)$, $(a,0)$ and radii r, R respectively, where $0 < r < R < 2a < r + R$. Then $\lambda = r + 2a - R$ and $\mu = R + 2a - r$. Hence in this case $h(S,T) = \mu$, as may be seen readily by reference to Figure 1. It may be shown that h is a complete metric on $\mathcal{C}(M)$, and that every closed bounded subset of $\mathcal{C}(M)$ is compact.

The definition is little more than a formal expression of the commonsense idea that two nonempty compact subsets S and T of M are, in an intuitive sense, 'close' to one another iff each point of S is close to some point of T, and each point of T is close to some point of S.

Figure 1. Hausdorff distance

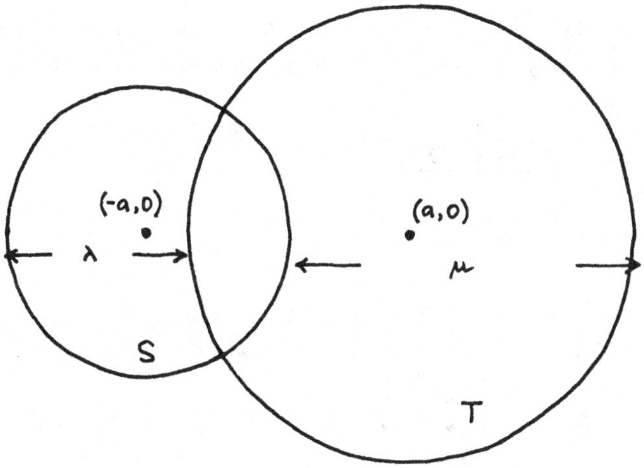

6. The space of polytopes

Let us apply the above construction to the case in which $(M,\delta) = (E,h)$. This yields a complete metric h on the set $\mathcal{C}(E) = \mathcal{C}$ of all nonempty compact subsets of E . If, therefore, we put $\mathcal{F}' = \mathcal{F} \cap \mathcal{C} = \mathcal{F} \setminus \{\emptyset\}$, and $\mathcal{P}' = \mathcal{P} \cap \mathcal{C} = \mathcal{P} \setminus \{\emptyset\}$, then $\mathcal{F}'$ and $\mathcal{P}'$ are subsets of $\mathcal{C}$, and we may give them the induced metrics $h_{\mathcal{F}}$ and $h_{\mathcal{P}}$ respectively and hence the corresponding metric topologies. To each space $\mathcal{F}'$ and $\mathcal{P}'$ we append $\emptyset = \{\emptyset\}$ as a singleton connected component, and so both $\mathcal{F}$ and $\mathcal{P}$ have been topologised.

It is easy to show that each of $\mathcal{F}$ and $\mathcal{P}$ has just two path-components. That is to say, both $\mathcal{F}'$ and $\mathcal{P}'$ are path-connected. For example, let $A, B \in \mathcal{F}'$ with $\# A = r$ and $\# B = s$. Let $C \in \mathcal{F}'$ be any set such that $\# C = rs$. Choose any labellings $a_1, \dots , a_r \in A$, $b_1, \dots , b_s \in B$, $c_{11}, c_{12}, \dots , c_{rs} \in C$ for the elements of A , B and C . For each pair (i,j) $(i=1,\dots ,r;\ j = 1,\dots ,s)$ define a path $\alpha_{ij}: I \to E$ by $\alpha_{ij}(t) = ta_i + (1-t)c_{ij}$, and let $\alpha: I \to \mathcal{F}'$ be given by $\alpha(t) = \{\alpha_{ij}(t): i=1,\dots ,r;\ j=1,\dots ,s\}$, where I denotes the closed interval $[0,1]$. Then α is a continuous path in $\mathcal{F}'$ from C to A . Likewise there is a path β from C to B and hence a path $\beta\alpha^{-1}$ from A to B in $\mathcal{F}'$. Figure 2 illustrates various steps along such a path for $r = 2$, $s = 3$.

Figure 2. Path-connectedness of $\mathcal{F}'$

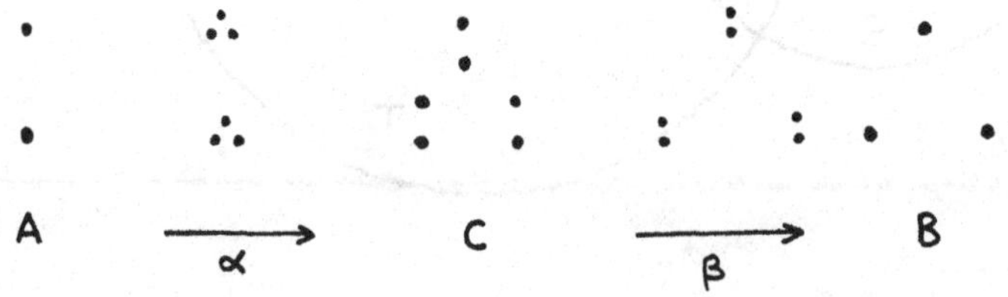

It is natural to ask whether the maps vert and conv are continuous with respect to the topologies that we have assigned to $\mathcal{P}$ and $\mathcal{F}$. We observe first that conv is continuous; for if A, B $\in \mathcal{F}'$, P = conv A and Q = conv B, then $h_{\mathcal{P}}(P,Q) \leq h_{\mathcal{F}}(A,B)$. On the other hand, vert is not continuous. To see this, let us construct a sequence (P_n) of polytopes P_n that converges in $\mathcal{P}$ to P but for which the corresponding sequence (V_n), where V_n = vert P_n, does not converge in $\mathcal{F}$ to V = vert P.

Consider the sequence of triangles P_n, where vert P_n = $\{(-1,0), (0,{}^1/_n), (1,0)\} \subset E^2$. Then (P_n) converges to the closed line-segment P = $[-1,1]$ in E^1, and so V = $\{-1,1\} = \{(-1,0),(1,0)\}$, while (V_n) converges to $\{(-1,0),(0,0),(1,0)\} = \{-1,0,1\}$. Figure 3 may make this a little more obvious.

Figure 3. vert is not continuous

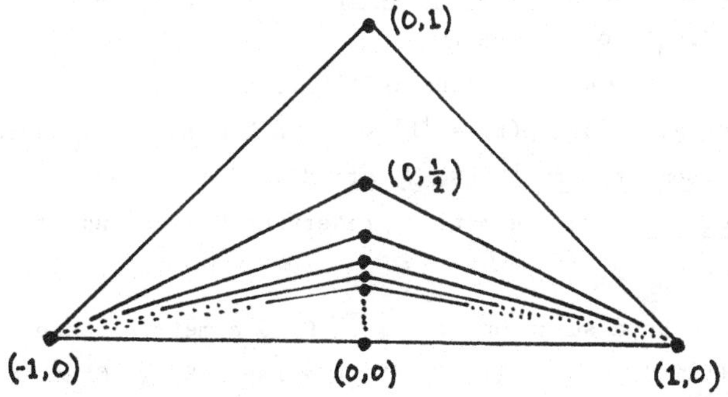

7. Similarity and congruence

In Euclid's Elements [Heath 1956], two equivalence relations between geometrical figures are much in evidence, either implicitly or explicitly, particularly in Euclid's treatment of triangles. Both relations may be defined in an obvious way on the set of subsets of any metric space, and have the virtue of being definable by means of group actions. Let us look first, then, at similarity and congruence in metric spaces.

Let (M,δ_M) and (N,δ_N) be metric spaces. A map $f:M \to N$ is said to be a similarity iff, for some real number $\lambda > 0$ and for all $x, y \in M$,

$$\lambda\delta_M(x,y) = \delta_N(f(x),f(y)) .$$

The number λ may be called the modulus of f, and we write accordingly $\lambda = |f|$. Let us denote the set of all similarities from M to N by $\mathcal{S}(M,N)$. If $f \in \mathcal{S}(M,N)$ and, for some metric space (W,δ_W), $g \in \mathcal{S}(N,W)$, then $g\circ f \in \mathcal{S}(M,W)$ and $|g\circ f| = |g||f|$. Trivially, $1_M \in \mathcal{S}(M,M)$ and we have, it seems, unwittingly constructed a category $\mathcal{S}$ of similarities between metric spaces. In this category $\mathcal{S}$, we are interested in the group $\mathrm{Aut}_{\mathcal{S}} M$ of invertible self-similarities of (M,δ_M). We call this the similarity group of (M,δ_M) and we prefer to denote it by $\mathrm{Sim}(M,\delta_M)$ or $\mathrm{Sim}\, M$.

The map $\mu:\mathrm{Sim}\, M \to R_*$ into the multiplicative group R_* of positive real numbers, given by $\mu(f) = |f|$, is a homomorphism whose kernel is called the isometry group $\mathrm{Iso}\, M$ of M. The elements of $\mathrm{Iso}\, M$, called isometries, are the metric-preserving bijections of M to itself.

Problems: (i) Let D be a subgroup of R_*. Find a metric space (M,δ_M) such that $\mu(\mathrm{Sim}\, M) = D$. (ii) Let $\Delta = \mathrm{im}\mu < R_*$. Characterise those metric spaces (M,δ_M) for which the short exact sequence

$$1 \longrightarrow \mathrm{Iso}\, M \longrightarrow \mathrm{Sim}\, M \xrightarrow{\mu} \Delta \longrightarrow 1$$

splits. Characterise those spaces (M,δ_M) for which $\Delta = R_*$.

The group $\mathrm{Sim}\, M$ acts on the set 2^M of all subsets of M by $f\cdot X = \{f(x):x \in X\}$, and $\mathrm{Iso}\, M$ also acts on 2^M by restriction of this action. An orbit of $\mathrm{Sim}\, M$ in 2^M is called a similarity class of subsets of M, and an orbit of $\mathrm{Iso}\, M$ is called a congruence class of subsets of M . In alternative language, if $X, Y \in 2^M$, then X is similar to Y , written $X \sim Y$, iff $Y = f \cdot X$ for some $f \in \text{Sim } M$, and X is congruent to Y , written $X \equiv Y$, iff $Y = g \cdot X$ for some $g \in \text{Iso } M$. Clearly $X \equiv Y$ implies $X \sim Y$, so that congruence is finer than similarity (and similarity is coarser than congruence).

We have approached these relations by using properties of the ambient space M as a whole. Another way to proceed is as follows. For $X \in 2^M$, let δ_X denote the metric induced on X by restriction of δ_M . Then we may choose to define a 'similarity' from (X,δ_X) to (Y,δ_Y) to be an $\mathcal{S}$-isomorphism in the categorical sense discussed above, and to say that X is 'similar' to Y iff such an $\mathcal{S}$-isomorphism exists. If $X \sim Y$, then X is certainly 'similar' to Y in this new sense. It may be impossible, however, to extend a given $\mathcal{S}$-isomorphism from (X,δ_X) to (Y,δ_Y) to obtain an element $f \in \text{Sim } M$, so the two approaches do not yield the same concept. Analogous remarks apply to congruence. For more details and references to the background literature, see Robertson [1976].

In the Euclidean spaces that we are studying here, these difficulties do not arise: the two approaches to similarity and congruence lead to the same pair of relations.

8. Euclidean similarity

Let us now consider the case $(M,\delta_M) = (E^n,d)$, where d denotes $d|E^n$. We put $\text{Sim } E^n = \text{Sim}(n)$, $\text{Iso}(E^n) = \text{Iso}(n)$, and observe that μ is surjective and the sequence

$$1 \longrightarrow \text{Iso}(n) \longrightarrow \text{Sim}(n) \xrightarrow{\mu} R_* \longrightarrow 1$$

splits (see Problems, §7). Thus Sim(n) may be identified with a semi-direct product of Iso(n) with R_* . As a set, therefore, we may identify Sim(n) with the Cartesian product of the underlying sets of Iso(n) and R_* . The group structure may be established once we know something about Iso(n) , which itself may be expressed as a semi-direct product.

Let R^n denote real linear n-space, and O(n) the orthogonal group of real $n \times n$ matrices H such that $HH^t = I_n$, where $I_n = [\delta_{ij}]$ is the $n \times n$ unit matrix and the superscript denotes transposition. Then there is a split short exact sequence

$$1 \longrightarrow R^n \longrightarrow \mathrm{Iso}(n) \longrightarrow O(n) \longrightarrow 1$$

under which we may identify Iso(n) as a set with the Cartesian product $R^n \times O(n)$, defining a group operation (juxtaposition) by

$$(a,H)(b,K) = (H\cdot b+a, HK) ,$$

where $\cdot$ denotes the standard action of O(n) on R^n.

We may then identify Sim(n) with the set $R^n \times O(n) \times R_*$ on which a group operation (again indicated by juxtaposition) is defined by

$$(a,H,s)(b,K,t) = (sH\cdot b+a, HK, st) .$$

The action of Sim(n) on E^n is given by

$$(a,H,s)\cdot x = sH\cdot x + a .$$

The next step is to transfer these ideas from E^n to E itself. The inclusion $E^n \subset E^{n+1}$ induces a monomorphism of Sim(n) into Sim(n+1), sending $(a,H,s) \in \mathrm{Sim}(n)$ to $(a,H',s) \in \mathrm{Sim}(n+1)$, where

$$H' = \begin{bmatrix} H & 0 \\ 0 & 1 \end{bmatrix} .$$

We may construct a group Sim as the direct limit of the resulting sequence of monomorphisms, and we identify Sim with the group constructed as follows. Let I_∞ denote the doubly-infinite matrix whose (i,j)th element is δ_{ij}, and let $\mathcal{O}$ denote the group of all doubly-infinite matrices of the form

$$\tilde{H} = \begin{bmatrix} H & 0 \\ 0 & I_\infty \end{bmatrix},$$

where $H \in O(n)$ for some n. Then $\mathrm{Sim} = E \times \mathcal{O} \times R_*$, with the group operation given by

$$(a,\tilde{H},s)(b,\tilde{K},t) = (s\tilde{H}\cdot b+a, \tilde{H}\tilde{K}, st)$$

as above, and with $(a,\tilde{H},s)\cdot x = s\tilde{H}\cdot x+a$ for $x \in E$.

The group Sim is a proper subgroup of Sim E. For consider the doubly infinite matrix

$$H_* = \mathrm{diag}(H,H,\ldots ,H,\ldots) ,$$

for any $H \in O(n)$. Then H_* acts on E by $H_*\cdot x = y$, where

$$y_{kn+i} = h_{ij}\, x_{kn+j} \quad (k=0,1,\ldots;\ i,j=1,\ldots,n) .$$

Thus $H_* \in \mathrm{Sim}\, E$ but $H_* \notin \mathrm{Sim}$. However, Sim is quite large enough for our purposes. In particular, the action of Sim E on 2^E leaves both $\mathcal{P}$ and $\mathfrak{F}$ setwise invariant. Moreover, if $P, Q \in \mathcal{P}$ are $\mathcal{S}$-isomorphic, then there exists $f \in \mathrm{Sim}$ such that $f \cdot P = Q$, and the same applies to $\mathfrak{F}$ and to $\mathcal{A}$. In fact, the maps conv, vert and aff are equivariant under the action of Sim.

In analogous fashion, we can construct a group $\mathrm{Iso} < \mathrm{Sim}$ and use this to define congruence in $\mathcal{P}$ and $\mathfrak{F}$.

9. The similarity space of polytopes

We are interested in polytopes with regard to their metrical symmetry. From this point of view, what matters is the shape of a polytope, and not its size or its position in space. A theory of symmetry for polytopes, therefore, must be a theory of similarity classes of polytopes rather than of polytopes as individuals. It follows that we should study the quotient space $\mathcal{P}/\sim \,= \mathcal{P}/\mathrm{Sim}$ rather than $\mathcal{P}$ itself.

For reasons that we now try to explain, we give $\mathcal{P}/\sim$ a modified version of the quotient topology. Our aim is to ensure that $\mathcal{P}/\sim$ is a Hausdorff space, retaining as far as possible the natural features of the Hausdorff metric topology on $\mathcal{P}$. The only neighbourhood of the class Θ of singleton subsets in $/\sim$ is $'/\sim$ itself, since any polytope is similar to one that is arbitrarily small. Accordingly, we consider the set $\mathcal{P}'$ as the union of the set $\mathcal{P}^o$ of polytopes of dimension 0 and the set $\mathcal{P}^+$ of polytopes of positive dimension. Thus $\mathcal{P} = \mathcal{P}^+ \cup \mathcal{P}^o \cup \{\emptyset\}$. We topologise $\mathcal{P}^+/\sim$ by the quotient topology, and attach $\Theta = \mathcal{P}^o/\sim$ and $\emptyset = \{\emptyset\} = \{\emptyset\}/\sim$ as two singleton path-components. The resulting space $\mathfrak{S}$ has three path-components, and is called the <u>similarity space</u> of polytopes. It is convenient to put $\mathfrak{S}^+ = \mathcal{P}^+/\sim$. Thus $\mathfrak{S}^+$ is the similarity space of polytopes of positive dimension.

One of the advantages of proceeding in this seemingly over-pedantic way lies in the fact that every element of $\mathfrak{S}^+$ includes 'normal' representatives, defined in §3.3 below.

We denote the similarity class of $P \in \mathcal{P}$ by $\$P$, and the congruence class of P by $\mathbb{K}P$.

2. COMBINATORIAL STRUCTURE

This discussion of the combinatorial structure of polytopes concentrates on concepts and theorems that are needed in Chapter 3. For more detailed treatment of standard ideas and results, and for developments of the subject in many other directions, see, for example, Grünbaum [1967] and Shephard & McMullen [1971].

I have retained established terminology as far as possible. Here and there, however, I have found it useful to introduce new terms such as 'multiplicity'. I have also preferred to employ nonstandard symbols for the members of the three standard sequences of regular polytopes.

1. Facial structure

Let P be an n-polytope with vert $P = V = \{v_1, \ldots, v_r\}$. For any subset W of V, $Q = \text{conv } W$ is an m-polytope for some $m \leq n$, and vert $Q = W$. Moreover $Q \subset P$. Any such polytope Q is called a subpolytope of P, written $Q < P$. Thus for any $P, Q \in \mathcal{P}$, $Q < P$ iff vert $Q \subset$ vert P. We say that Q is a k-subpolytope of P iff $Q < P$ and $\dim Q = k$.

We now pick out certain special subpolytopes of P called the faces of P, as follows. Let $X \subset V$, and let $X' = V \setminus X$. Consider the polytopes $Q = \text{conv } X$, $Q' = \text{conv } X'$ and the affine plane $K = \text{aff } X$. If $K \cap Q' = \emptyset$, then Q is said to be a face of P, and to be a k-face of P if, furthermore, $\dim Q = k$. We write $Q \triangleleft P$ to mean that Q is a face of P. We say that Q is incident with P iff $Q < P$ and $\dim Q < \dim P$. Of course $P \triangleleft P$ and $\emptyset \triangleleft P$, for all P. In fact, P is the unique n-subpolytope of the n-polytope P, and $\emptyset$ is its unique (-1)-subpolytope. Figure 1 illustrates some of these ideas in the very simple case where P is a square. For every vertex v of P, $\{v\}$ is a 0-face of P. Usually, for simplicity, we do not distinguish between v and $\{v\}$, and refer to v either as a vertex or a 0-face. The 1-faces of P are called its edges and the (n-1)-faces its facets (at least for $n \geq 4$). The facets of a 3-polytope or polyhedron are, by long-established custom, called its faces.

The two relations $<$ and $\triangleleft$ yield posets $(\mathcal{P}, <)$ and $(\mathcal{P}, \triangleleft)$, and $\emptyset$ is the minimal element in both. For any $P \in \mathcal{P}$, we consider the set $S(P) = \{Q \in \mathcal{P} : Q < P\}$ of subpolytopes of P and the set $F(P) =$

$\{Q \in \mathcal{P} : Q \lhd P\}$ of faces of P. Each is a lattice with respect to its partial ordering, with maximal element P and minimal element $\emptyset$, and of course $(F(P), \lhd)$ is a sublattice of $(S(P), <)$. We denote the sets of j-subpolytopes and of j-faces of P by $S_j(P)$ and $F_j(P)$ respectively, $j = -1, 0, \ldots, n = \dim P$. Note that $S_j(P) \supset F_j(P)$ for all j.

Among the basic facts about the structure of $F(P)$, the following are particularly important, but we omit the proofs since they are readily available elsewhere (for example, Grünbaum [1967]).

Exercise: Show that for any n-polytope P, where $n \geq 2$, $F_j(P) \neq \emptyset$ for all $j = -1, \ldots, n$ and for each vertex v and each facet T of P there is a sequence

$$v = T_0 \lhd T_1 \lhd \ldots \lhd T_{n-1} = T,$$

where $T_i \in F_i(P)$, $i = 0, 1, \ldots, n-1$.

Exercise: Show that for any n-polytope P, where $n \geq 2$, for each $T \in F_{n-1}(P)$ there are exactly two $(n-2)$-faces T' and T'' such that $T' \lhd T$ and $T'' \lhd T$.

Figure 1. Subpolytopes and faces of a square

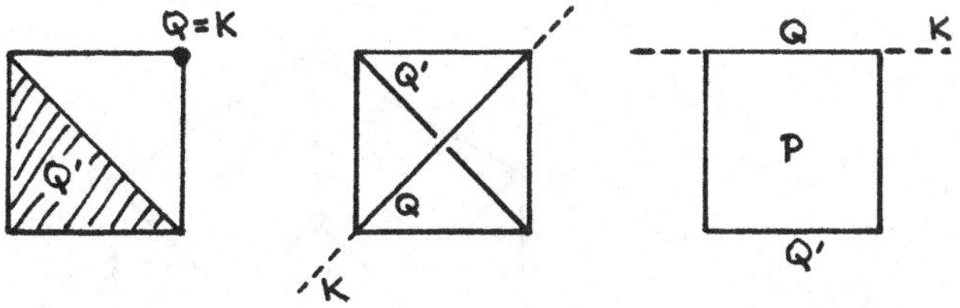

(i) $Q \lhd P$ (ii) $Q < P$, $Q \ntriangleleft P$ (iii) $Q \lhd P$

2. Combinatorial equivalence

In the combinatorial theory of polytopes, the lattice $(F(P),\triangleleft)$, called the <u>face-lattice</u> of P, and usually denoted simply by $F(P)$, is regarded as the main structural feature of P. From this point of view, the fundamental equivalence relation between polytopes is <u>combinatorial equivalence</u>, denoted by $\approx$ and defined by putting $P \approx Q$ iff there is a lattice-isomorphism $\lambda : F(P) \to F(Q)$. For example, any two triangles are combinatorially equivalent, since the face-lattice of any triangle is isomorphic to that shown schematically in Figure 2 (with 'traditional' labelling of vertices and edges). Any combinatorial equivalence $\lambda : F(P) \to F(Q)$ induces a bijection $\lambda_j : F_j(P) \to F_j(Q)$, $j = 1,\ldots,s$, and λ is determined by λ_0. Any such λ_0 is called a <u>combinatorial isomorphism</u>.

It may be noted that $P \approx Q$ does not imply that $(S(P),<)$ is lattice-isomorphic to $(S(Q),<)$. To see this, consider the octahedra P with vertices $(\pm1,0,0)$, $(0,\pm1,0)$ and $(0,0,\pm1)$ and Q with vertices $(\pm1,0,\varepsilon)$, $(0,\pm1,-\varepsilon)$ and $(0,0,\pm1)$ in E^3, where $\varepsilon > 0$ is small. Then $P \approx Q$ but $S(P)$ is not isomorphic to $S(Q)$. I am not aware that any attempt has been made to develop a theory based on isomorphism of the subpolytope lattices.

Figure 2. <u>A triangle and its face-lattice</u>

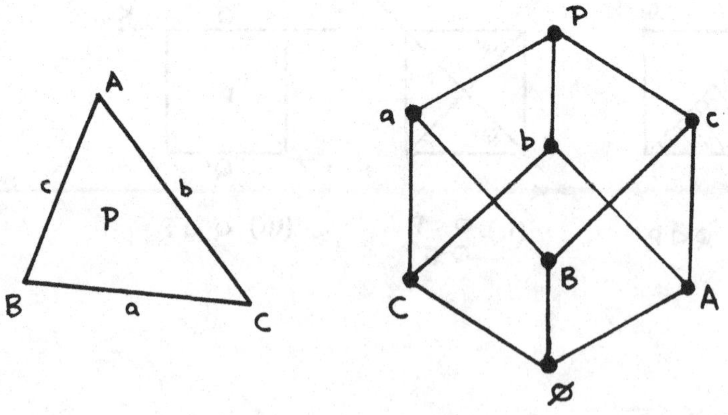

The combinatorial equivalence class to which a polytope P belongs is denoted by $\mathbb{C}P$ and is called the combinatorial type of P. The first obvious problem is to devise combinatorial invariants, that is to say maps with domain $\mathcal{P}$ that are constant on combinatorial types. Of particular interest are complete combinatorial invariants, which are invariants inducing injective maps on the set of combinatorial types. Thus in the diagram

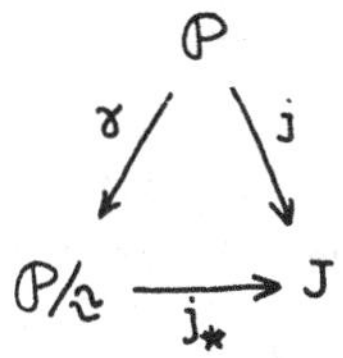

$j:\mathcal{P}\to J$ is a combinatorial invariant if $j_*:\mathcal{P}/\approx\ \to J$ exists such that $j_*\circ\gamma = j$, where $\gamma:\mathcal{P}\to\mathcal{P}/\approx$ is the quotient map, and such an invariant is complete iff j_* is injective.

The isomorphism class of F(P) is, by definition, a complete combinatorial invariant. Among other (mostly incomplete) invariants, we mention the following.

Let $f_i(P) = \# F_i(P)$. Then $f(P) = (f_0(P), \ldots, f_{n-1}(P))$, where $n = \dim P$, is called the face-vector of P. Clearly f(P) and its components $f_i(P)$, $i = 0, \ldots, n-1$, are combinatorial invariants. A glance at Figure 3, however shows that f(P) is incomplete, since the two polyhedra shown have the same face-vector but are not combinatorially equivalent. The components of the face-vector are not independent of one another. In particular, they satisfy the Euler relation $\sum_{i=-1}^{n} (-1)^{i-1} f_i(P) = 0$, where $n = \dim P$. (See, for example, Grünbaum [1967] Chapter 8.)

The group Aut P of all lattice-automorphisms of F(P) is a useful invariant, but again Aut P is incomplete. For instance, it is easy to devise examples of polytopes P in any dimension $n \geq 3$ for which Aut P is trivial.

An interesting invariant that turns out to be complete is the incidence matrix, defined as follows. Let $v_1, \ldots, v_r$ denote the vertices of P and $T_1, \ldots, T_s$ its facets. Consider the matrix $M(P)$ of order $r \times s$ whose (i,j)th element $\mu_{ij}(P)$ is 1 if $v_i \triangleleft T_j$ and is 0 otherwise. Then $M(P)$ is a complete combinatorial invariant in the sense that if P and Q are polytopes, then $P \approx Q$ iff , for some labellings of vertices and facets, $M(P) = M(Q)$. To see that this is so, we have to satisfy ourselves that the structure of $F(P)$ can be deduced from a knowledge of $M(P)$. Now any two facets of P meet in a unique $(n-2)$-face , and each $(n-2)$-face of P lies in exactly two facets. The $(n-2)$-faces of P can be read off from $M(P)$ as follows.

Let T_a and T_b be facets of P . Consider the subset $W = \{w_1, \ldots, w_t\} = \{v_{k_1}, \ldots, v_{k_t}\}$ of vert P such that $w_i \triangleleft T_a$ and $w_i \triangleleft T_b$ for all $i = 1, \ldots, t$. Then $S = \text{conv } W$ is an $(n-2)$-face of P iff T_a and T_b are the only facets such that $S \triangleleft T_a$ and $S \triangleleft T_b$, and we can check whether this is so by inspection of $M(P)$. Every such S is a facet of T_a . In this way we can determine the set $F_{n-2}(P)$ and the incidences of $F_{n-2}(P)$ with $F_{n-1}(P)$.

We can now construct an incidence matrix for each facet T_a of P by writing out which vertices of T_a lie in each $(n-2)$-face of the $(n-1)$-polytope T_a . By iterating this procedure, we may establish the full structure of $F(P)$, represented as a lattice of subsets of $F_0(P)$.

Figure 3. $f(P) = f(Q) \not\Rightarrow P \approx Q$

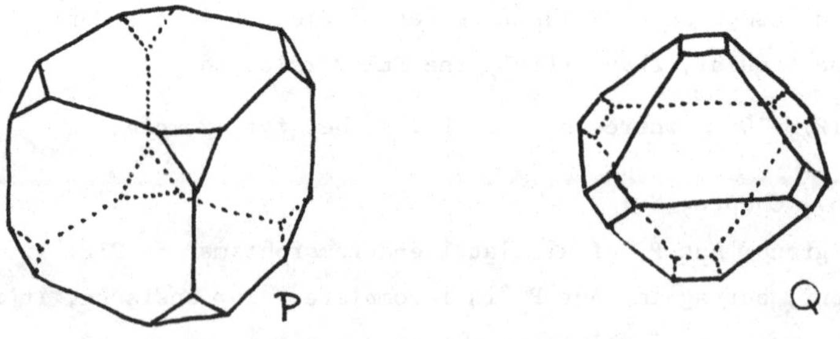

$f(P) = (24,36,14) = f(Q)$

Let us take a simple example. Consider a pyramid P with quadrilateral base, labelling the vertices and face(t)s as shown in Figure 4. Then

$$M(P) = \begin{bmatrix} 0 & 1 & 1 & 1 & 1 \\ 1 & 0 & 1 & 1 & 0 \\ 1 & 0 & 0 & 1 & 1 \\ 1 & 1 & 0 & 0 & 1 \\ 1 & 1 & 1 & 0 & 0 \end{bmatrix} .$$

For any $a, b \in \{1,2,3,4,5\}$, $\mu_{ia} = \mu_{ib} = 1$ for at most two values of $i \in \{1,2,3,4,5\}$, and the pairs (a,b) , $a < b$, for which two such values of i exist are the pairs of faces of P meeting in an edge. The edges of P are then easily listed, together with the pairs of faces to which they belong. Thus there is an edge with vertices v_4 and v_5 common to the faces T_1 and T_2 , since columns 1 and 2 of $M(P)$ both have entry 1 in rows 4 and 5 .

An auxiliary invariant that we shall need shortly is the number of nonzero elements (i.e. the number of entries equal to 1) in $M(P)$. We call this number $\mu(P)$ the <u>multiplicity</u> of P . Thus $\mu(P)$ is the number of pairs $(v,T) \in F_0(P) \times F_{n-1}(P)$ with $v \lhd T$. For the above pyramid P , $\mu(P) = 16$.

Figure 4. <u>A labelled pyramid</u>

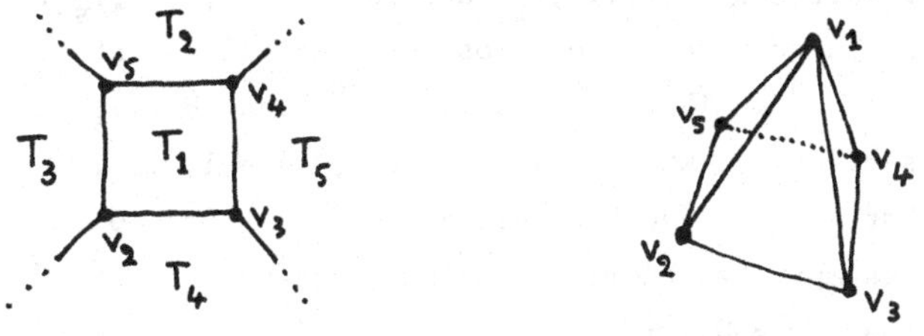

Exercise: Let P be an n-polytope. For each $k, \ell \in \{0,1,\ldots,n-1\}$, with $k < \ell$, let $I_{k,\ell}(P)$ denote the matrix with (i,j)th element $m_{ij} = 1$ if $A_i < B_j$ and $m_{ij} = 0$ otherwise, where $F_k(P) = \{A_1,\ldots,A_r\}$, $F_\ell(P) = \{B_1,\ldots,B_s\}$, $r = f_k(P)$, $s = f_\ell(P)$. In particular, let $J_k = I_{k,k+1}(P)$. Calculate J_0 and J_1 for the pyramid P discussed above, choosing some labelling for its edges. Check that $J_0J_1 = 0$. It is true generally that $J_kJ_{k+1} = 0$, this being a manifestation of the boundary operator relation $\partial_{k+1}\partial_k = 0$ familiar in elementary homology theory. See Coxeter [1948].

3. Topological structure of combinatorial types

Let P be an n-polytope. Then $\mathbb{C}P$ is a subset of $\mathcal{P}$. What can be said about the topological structure of $\mathbb{C}P$? We try to answer this question by looking at $\mathbb{C}P$ as the union of the <u>sub-types</u> $\mathbb{C}_mP$, $m \geq n$, where $\mathbb{C}_mP$ denotes the set of all $Q \in \mathcal{P}_m$ such that $Q \approx P$. Thus $\mathbb{C}_mP = \mathbb{C}P \cap \mathcal{P}_m$. We call $\mathbb{C}_nP = \mathbb{C}_*(P)$ the <u>principal</u> sub-type of P, and concentrate first of all on the structure of this space.

Theorem: <u>Let P be an n-polytope. Then $\mathbb{C}_*(P)$ has the structure of a topological manifold of dimension</u>

$$n(f_0(P) + f_{n-1}(P)) - \mu(P) .$$

Proof: The proof rests on the fact that the incidence matrix is a complete combinatorial invariant, and is built round a simple application of the implicit function theorem.

Let P have vertices $v_1,\ldots,v_r$ and facets $T_1,\ldots,T_s$. Since $\dim P = n$, we can suppose without loss of generality that $P \in C_*P$, i.e. that $P \subset E^n$. Thus $v_1,\ldots,v_r \in E^n$ and $H_1,\ldots,H_s$ are affine hyperplanes of E^n, where $H_j = \text{aff } T_j$, $j = 1,\ldots,s$. We can also suppose that none of the hyperplanes H_j contains 0.

With these assumptions, there are unique nonzero elements $a_1,\ldots a_s \in E^n$ such that $x \in H_j$ iff $\langle x,a_j\rangle = 1$, $j = 1,\ldots,s$. The combinatorial structure of P is now specified completely by the $\mu(P)$ conditions

$$\langle v_i,a_j\rangle = 1 \quad \text{iff} \quad \mu_{ij}(P) = 1 .$$

Suppose now that Q is an n-polytope in E^n such that $Q \approx P$ and which is 'close' to P in the sense that its vertices are $w_1, \ldots, w_r$ where, for some $\varepsilon < 0$ and all $i = 1, \ldots, r$, $\|w_i - v_i\| \leq \varepsilon$ and the map $\beta: F_0(P) \to F_0(Q)$ given by $\beta(v_i) = w_i$ is a combinatorial isomorphism. Let the combinatorial equivalence $\lambda: F(P) \to F(Q)$ determined by β be such that $\lambda(A_j) = B_j$, for some labelling of $F_{n-1}(Q)$. Now each hyperplane $K_j = \text{aff } B_j$ is given by a unique equation $\langle x, b_j \rangle = 1$, provided ε is sufficiently small, where $b_j \neq 0$.

Let $\xi_i = w_i - v_i$ and $\eta_j = b_j - a_j$. Since

$$\langle w_i, b_j \rangle = 1 \quad \text{iff } \mu_{ij}(Q) = 1 ,$$

and $\mu_{ij}(P) = \mu_{ij}(Q)$ for all i, j, it follows that

$$\langle v_i + \xi_i , a_j + \eta_j \rangle = 1 \quad \text{iff } \mu_{ij}(P) = 1 ,$$

and hence

$$\langle \xi_i, a_j \rangle + \langle v_i, \eta_j \rangle + \langle \xi_i, \eta_j \rangle = 0 \quad \text{iff } \mu_{ij}(P) = 1 .$$

These observations suggest that we consider the polynomial map $\Phi: R^{nr} \times R^{ns} \to R^m$ of degree 2 in the components of $(\xi_1, \ldots, \xi_r) \varepsilon (R^n)^r = R^{nr}$ and $(\eta_1, \ldots, \eta_s) \varepsilon (R^n)^s = R^{ns}$, where $m = \mu(P)$, defined as follows.

Let the pairs (i,j) such that $\mu_{ij}(P) = 1$ be put in lexicographical order, and (i,j) occupy the $[i,j]$th in this list. Put $\Phi(\xi, \eta) = \zeta$, where $\xi = (\xi_1, \ldots, \xi_r)$, $\eta = (\eta_1, \ldots, \eta_s)$, $\zeta = (\zeta_1, \ldots, \zeta_m)$, and for all $[i,j] = 1, \ldots, m$,

$$\zeta_{[i,j]} = \langle \xi_i, a_j \rangle + \langle v_i, \eta_j \rangle + \langle \xi_i, \eta_j \rangle .$$

For some open neighbourhood N of $(0,0)$ in $R^{nr} \times R^{ns}$, there is a homeomorphism $h: N_0 \to N_P$ from $N_0 = N \cap \Phi^{-1}(0)$ to some open neighbourhood N_P of P in $\mathbb{C}_* P$, given by $h(\xi, \eta) = Q$, where the vertices of Q are $v_1 + \xi_i$ and its facets are given as above by $a_j + \eta_j$.

To prove the Theorem, we have only to show that the map Φ has the rank m at $(0,0)$. We must therefore calculate the Jacobian matrix $J\Phi(0,0)$ of the map Φ at $(0,0)$. We observe that, with an obvious notational convention in which sequences of real variables are 'compressed' into vectors,

$$\frac{\partial \zeta_{[i,j]}}{\partial \xi_i} = a_j + \eta_j ,$$

and

$$\frac{\partial \zeta_{[i,j]}}{\partial \eta_j} = v_i + \xi_i ,$$

for all (i,j) such that $\mu_{ij}(P) = 1$. Moreover, all other entries of $J\Phi(\xi,\eta)$ vanish. It follows that $J = J\Phi(0,0)$ has a_j in the ith column and $[i,j]$th row, v_i in the jth column and $[i,j]$th row, and all other entries are zero.

Suppose then that some linear combination of the rows of J is the zero vector in $R^{n(r+s)}$, and let λ_{ij} be the coefficient of the $[i,j]$th row. Then the m numbers λ_{ij} satisfy the equations

(1) $\quad \sum_j \lambda_{ij} a_j = 0 , \qquad i = 1,\dots,r ,$

and

(2) $\quad \sum_i \lambda_{ij} v_i = 0 , \qquad j = 1,\dots,s ,$

the summations being over all $[i,j]$, for fixed i and j respectively. Since, for all $[i,j]$,

(3) $\quad \langle v_i , a_j \rangle = 1 ,$

we also observe that

(4) $\quad \sum_j \lambda_{ij} = 0 , \qquad i = 1,\dots,r ,$

and

(5) $\quad \sum_i \lambda_{ij} = 0 , \qquad j = 1,\dots,s .$

Now the matrix of coordinates $(a_j)_p$ in (1) has rank n, since the facets T_j meet in the unique point v_i. Likewise, the matrix of coordinates $(v_i)_q$ in (2) has rank n, since the points v_i determine the hyperplane T_j. The solution space of (1) is therefore a linear space of dimension $s_i - n$, where s_i is the number of pairs $[i,j]$, and that of (2) is of dimension $r_j - n$, where r_j is the number of pair $[i,j]$. But λ_{ij} is the only coefficient appearing in both (1) and (2), for each $[i,j]$, so the solution space of the entire set of equations (1) and (2) is $m - (r+s)n$. But $m < (r+s)n$, since the polytope P is constructed by imposing m linear conditions (3) on the $(r+s)n$ coordinates $(v_i)_q$, $(a_j)_p$. It follows that $\lambda_{ij} = 0$ for all $[i,j]$. That is, J has rank m. Equations (4) and (5) are not needed.

The method of proof constructs an atlas of local coordinate systems or charts on the space $\mathbb{C}_*P$. These are the maps $h^{-1}:N_p \to N_O$

discussed in the proof. It is tempting to explore the nature of the transition homeomorphisms for pairs of charts with overlapping domains, showing that these maps are algebraic. However, we do not pursue this investigation here, although the questions raised may be of considerable interest. At the very least, $\mathbb{C}_m P$ is differentiable.

The reader may find it helpful to see the above ideas worked out for a particular polytope. Let us take the pyramid P of Figure 4. In this case $n = 3$, $r = s = 5$ and $m = 16$. So J is a matrix of order 16×30, or 16×10 in the 'compressed' notation. We find that

$$J = \begin{bmatrix}
a_2 & \cdot & \cdot & \cdot & \cdot & \cdot & v_1 & \cdot & \cdot & \cdot \\
a_3 & \cdot & \cdot & \cdot & \cdot & \cdot & \cdot & v_1 & \cdot & \cdot \\
a_4 & \cdot & \cdot & \cdot & \cdot & \cdot & \cdot & \cdot & v_1 & \cdot \\
a_5 & \cdot & \cdot & \cdot & \cdot & \cdot & \cdot & \cdot & \cdot & v_1 \\
\cdot & a_1 & \cdot & \cdot & \cdot & v_2 & \cdot & \cdot & \cdot & \cdot \\
\cdot & a_3 & \cdot & \cdot & \cdot & \cdot & \cdot & v_2 & \cdot & \cdot \\
\cdot & a_4 & \cdot & \cdot & \cdot & \cdot & \cdot & \cdot & v_2 & \cdot \\
\cdot & \cdot & a_1 & \cdot & \cdot & v_3 & \cdot & \cdot & \cdot & \cdot \\
\cdot & \cdot & a_4 & \cdot & \cdot & \cdot & \cdot & \cdot & v_3 & \cdot \\
\cdot & \cdot & a_5 & \cdot & \cdot & \cdot & \cdot & \cdot & \cdot & v_3 \\
\cdot & \cdot & \cdot & a_1 & \cdot & v_4 & \cdot & \cdot & \cdot & \cdot \\
\cdot & \cdot & \cdot & a_2 & \cdot & \cdot & v_4 & \cdot & \cdot & \cdot \\
\cdot & \cdot & \cdot & a_5 & \cdot & \cdot & \cdot & \cdot & \cdot & v_4 \\
\cdot & \cdot & \cdot & \cdot & a_1 & v_5 & \cdot & \cdot & \cdot & \cdot \\
\cdot & \cdot & \cdot & \cdot & a_2 & \cdot & v_5 & \cdot & \cdot & \cdot \\
\cdot & \cdot & \cdot & \cdot & a_3 & \cdot & \cdot & v_5 & \cdot & \cdot
\end{bmatrix}$$

The rank of J is 16, and $\mathbb{C}_* P$ is a topological manifold of dimension $3(5+5) - 16 = 14$. There is an alternative, intuitive, way of calculating the dimension of $\mathbb{C}_* P$ which we illustrate in the case of the pyramid as follows.

Let us picture the way in which P, or any polyhedron $Q \approx P$, can be constructed in E^3. The first vertex v_1 can be chosen anywhere in E^3 , and so has three degrees of freedom. The second vertex v_2 can then be chosen anywhere in $E^3 \setminus \{v_1\}$, and it too, therefore, has three degrees of freedom. Likewise, v_3 can be chosen anywhere in E^3 except on the line $v_1 v_2$. So v_3 has three degrees of freedom. Now v_4 can be chosen anywhere in E^3 except on the plane $v_1 v_2 v_3$, and so v_4 has three degrees of freedom. Having chosen v_1, v_2, v_3 and v_4 , v_5 must be chosen to lie on the plane $v_2 v_3 v_4$ and outside the triangle $v_2 v_3 v_4$. So v_5 has two degrees of freedom in this mode of construction. In all, we have $3 + 3 + 3 + 3 + 2 = 14$ degrees of freedom.

It is quite easy to deduce from the Theorem a formula for the dimension of $\mathbb{C}_m P$, for any $m \geq n$.

Corollary: <u>Let P be an n-polytope. Then for any $m \geq n$, $\mathbb{C}_m P$ is a topological manifold of dimension</u>

$$(m-n)(n+1) + n(f_0(P) + f_{n-1}(P)) - \mu(P) .$$

Proof: Let $Q \in \mathbb{C}_m P$. Then aff Q is an affine n-plane in E^m , and $\mathbb{C}_m P$ is the union of the disjoint families of n-polytopes $Q' \subset A$, where $Q' \in \mathbb{C}_m P$ and A runs through the family of all such affine n-planes in E^m . Thus $\mathbb{C}_m P$ is a fibre bundle over the Grassmann space $\mathcal{A}_{n,m}$ of all affine n-planes in E^m , with fibre $C_* P$. Since $\mathcal{A}_{n,m}$ is a manifold of dimension $(m-n)(n+1)$, the above formula is proved correct.

The above results help us to form a clearer picture of the space $\mathcal{P}_m$: it is partitioned into topological manifolds or <u>strata</u>, namely the combinatorial sub-types $\mathbb{C}_m P$, $P \in \mathcal{P}_m$; moreover, the frontier of each such stratum in $\mathcal{P}_m$ is composed of finitely many such strata, of lower dimension. For example, if $P \in \mathcal{P}_3$ is the pyramid discussed above, then the frontier of $\mathbb{C}_3 P$ in $\mathcal{P}_3$ is made up of six other strata: tetrahedra, pentagons, quadrilaterals, triangles, line-segments and singletons. These strata have dimensions 12, 13, 11, 9, 6 and 3 respectively.

The inclusion $\mathcal{P}_m \subset \mathcal{P}_{m+1}$ is stratum-preserving, and we can represent $\mathcal{P}$ itself as a space partitioned into strata of countably infinite dimension by taking direct limits of strata in the sequence $(\mathcal{P}_m)$.

4. The three standard sequences

We now introduce three standard sequences of polytopes that are important in both the combinatorial and the metrical theories. At this stage in our discussion, they serve to illustrate some of the ideas that have been considered in preceding sections. Their central rôle in the theory of polytopes may become clearer as the story unfolds.

Let n be an integer such that $n \geq 1$. The standard regular $(n-1)$-*simplex* Δ_{n-1} is the $(n-1)$-polytope in E^n whose vertices are the elements $e_1, \ldots, e_n$ of the standard basis of E^n. Thus $(e_i)_j = \delta_{ij} = 1$ if $i = j$ and $(e_i)_j = 0$ if $i \neq j$. For example, Δ_2 has vertices $(1,0,0)$, $(0,1,0)$ and $(0,0,1)$ in E^3, and is an *equilateral triangle*. The polyhedron Δ_3 is a *regular tetrahedron*.

The standard n-cube $\square_n$ is the n-polytope in E^n whose vertices are the 2^n points $(\pm 1, \ldots, \pm 1) = \pm e_1 \pm \ldots \pm e_n$. Thus $\square_2$ is the square in E^2 with vertices $(1,1)$, $(1,-1)$, $(-1,1)$ and $(-1,-1)$, while $\square_3$ is a *cube* or *regular hexahedron* in E^3.

The standard n-*cross-polytope* $\Diamond_n$ is the n-polytope in E^n whose vertices are the $2n$ points $\pm e_1, \ldots, \pm e_n$. Thus $\Diamond_2$ is a square in E^2 with vertices $(1,0)$, $(-1,0)$, $(0,1)$ and $(0,-1)$, while $\Diamond_3$ is a *regular octahedron*.

Figure 5. *Three standard polygons*

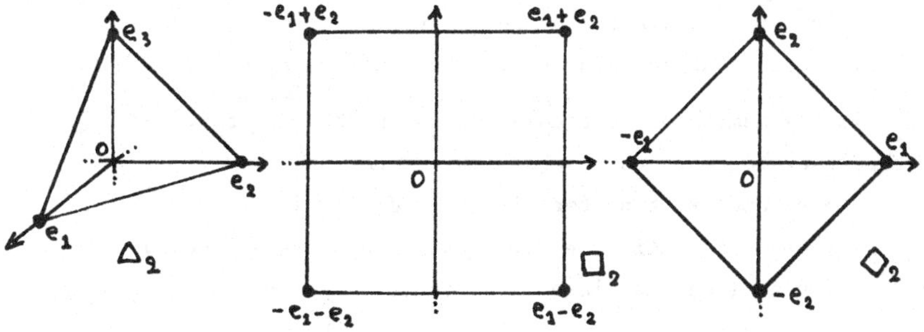

Note that $\Delta_n \not\approx \Box_n \not\approx \Diamond_n$ iff $n = 1$ and $\Box_n \not\approx \Diamond_n$ iff $n = 1$ or $n = 2$.

The various combinatorial invariants that we have defined can be easily calculated for these polytopes. Thus

$$f_i(\Delta_n) = \binom{n+1}{i+1}, \quad f_i(\Box_n) = 2^{n-i}\binom{n}{i}, \quad \text{and} \quad f_i(\Diamond_n) = 2^{i+1}\binom{n}{i+1},$$

where the brackets refer to binomial coefficients.
For example,

$$f(\Delta_3) = (4,6,4), \quad f(\Box_3) = (8,12,6), \quad f(\Diamond_3) = (6,12,8).$$

It may be helpful to remark that Coxeter [1948] denotes Δ_n by α_n, $\Box_n$ by γ_n and $\Diamond_n$ by β_n. He also calls $\Box_n$ a <u>measure polytope</u>.

Exercise: Show that

$$\mu(\Delta_n) = n(n+1), \quad \mu(\Box_n) = n2^n = \mu(\Diamond_n).$$

Exercise: Show that $\mathrm{Aut}\,\Box_n$ has order $n!2^n$, and construct an isomorphism from $\mathrm{Aut}\,\Delta_n$ to the symmetric group S_{n+1} of degree $n + 1$.

Exercise: Calculate $\dim \mathbb{C}_n P$ for $P = \Delta_n$, $\Box_n$ and $\Diamond_n$.

Exercise: Prove that if P is an n-polytope, then $f_0(P) \geq n + 1$ and that $P \approx \Delta_n$ if $f_0(P) = n + 1$.

5. Simple and simplicial polytopes

For any n-polytope P with vertices $v_1,\dots,v_r$ and facets $T_1,\dots,T_s$, let r_j denote the number of vertices in T_j and s_i the number of facets that contain v_i. Then $\sum_i s_i = \sum_j r_j = \mu(P)$. Of course s_i is the number of nonzero elements in the ith row, and r_j the number of nonzero elements in the jth column, of $M(P)$. Moreover, $s_i \geq n$ and $r_j \geq n$ for all i and all j.

If $r_j = n$ for all $j = 1,\dots,s$, then every facet of P is an (n-1)-simplex (see Exercises in §4 above), and P is said to be <u>simplicial</u>. If $s_i = n$ for all $i = 1,\dots,r$, then P is said to be <u>simple</u>.

The polytope Δ_n is both simple and simplicial, while $\Box_n$ is simple and $\Diamond_n$ is simplicial. Every polygon is both simple and simplicial.

Exercise: Prove that each of the sets of simple polytopes and of simplicial polytopes is dense in $\mathcal{P}$.

There are plenty of polytopes that are neither simple nor simplicial: the pyramid of §2 is an obvious example.

Exercise: Prove that if P is a simple n-polytope, then $\mu(P) = nf_0(P)$, and if P is simplicial, then $\mu(P) = nf_{n-1}(P)$. Deduce that if P is simple then $\dim \mathbb{C}_*P = nf_{n-1}(P)$, and if P is simplicial, then $\dim \mathbb{C}_*P = nf_0(P)$.

6. Duality and polarity

Recall that if P and Q are polytopes, then P is said to be combinatorially equivalent to Q , written $P \approx Q$, iff there is a lattice isomorphism $\lambda: F(P) \to F(Q)$ (or combinatorial equivalence from P to Q). The concept of duality is in one sense an 'inverted' form of this relation: we say that P is dual to Q , written $P \wr\wr Q$, iff there is an anti-isomorphism $\alpha: F(P) \to F(Q)$ (or duality equivalence from P to Q). That is to say, $P \wr\wr Q$ iff there is a bijection $\alpha: F(P) \to F(Q)$ such that, for all faces $S,T \in F(P)$, $S \lhd T$ iff $\alpha(T) \lhd \alpha(S)$.

Just as a combinatorial equivalence from P to Q is determined by its restriction to $F_0(P)$, called a combinatorial isomorphism, so a duality equivalence from P to Q is determined by its restriction to $F_0(P)$, called a dualisation from P to Q .

Whereas $\approx$ is an equivalence relation on $\mathcal{P}$, $\wr\wr$ is neither reflexive nor transitive. It is, however, a symmetric relation. Moreover, $\wr\wr$ is a relation between combinatorial types: for if $P \approx P'$ and $Q \approx Q'$, then $P \wr\wr Q \iff P' \wr\wr Q'$.

Although it is false in general that $P \wr\wr P$, there are many examples of polytopes that are self-dual in this sense. For example, our increasingly familiar friend the pyramid of §2 is self-dual. Likewise, $\Delta_n \wr\wr \Delta_n$ for all n , and $P \wr\wr P$ for every polygon P . It is an easy exercise to check that $\square_n \wr\wr \diamond_n$ for all n .

If $P \wr\wr Q$, then we should expect that the combinatorial invariants of Q are closely related to those of P . Thus if $\alpha: F(P) \to F(Q)$ is a duality equivalence from P to Q , and if P has vertices $v_1, \dots ,v_r$ and facets $T_1, \dots ,T_s$, while Q has vertices $w_1, \dots ,w_s$ and facets $S_1, \dots S_r$, where $\alpha(v_i) = S_i$ and $\alpha(T_j) = w_j$, then $M(Q)$ is the transpose of $M(Q)$ with respect to these labellings. Thus $\mu(P) = \mu(Q)$. Also $\mathrm{Aut}\, P \cong \mathrm{Aut}\, Q$, a particular isomorphism being given by mapping $g \in \mathrm{Aut}\, P$ to $\alpha \circ g \circ \alpha^{-1} \in \mathrm{Aut}\, Q$. If $\dim P = n$ ($= \dim Q$) , then $f_i(P) = f_{n-i-1}(Q)$ for all $i = 0, \ldots, n-1$. In particular, the face-vector of any self-dual polytope is palindromic: for instance, $f(\Delta_5) = (6,15,20,15,6)$.

We show next that for any polytope P there is a polytope Q such that $P \wr\wr Q$. Trivially, $P \wr\wr P$ if $\dim P < 2$. The following construction applies to n-polytopes P with $n \geq 2$. (of course $P \wr\wr P$ for any polygon.) Let P be an n-polytope in E^n with centre c , and let S_ρ denote the sphere with centre c and radius ρ , where $\rho > 0$. Suppose that P has vertices $v_1, \ldots, v_r$, and for each $i = 1, \ldots, r$ let $\tilde{v}_i$ denote the polar of v_i with respect to S_ρ . Then $\tilde{v}_i$ is an affine $(n-1)$-plane in E^n and $x \in \tilde{v}_i$ iff $\langle x-c, v_i-c \rangle = \rho^2$, as indicated in Figure 6. Now there is a unique n-polytope P_ρ^* in E^n whose facets $v_1^*, \ldots, v_r^*$ are such that $\mathrm{aff}\, v_i^* = \tilde{v}_i$.

Figure 6. <u>Vertex and polar hyperplane</u>

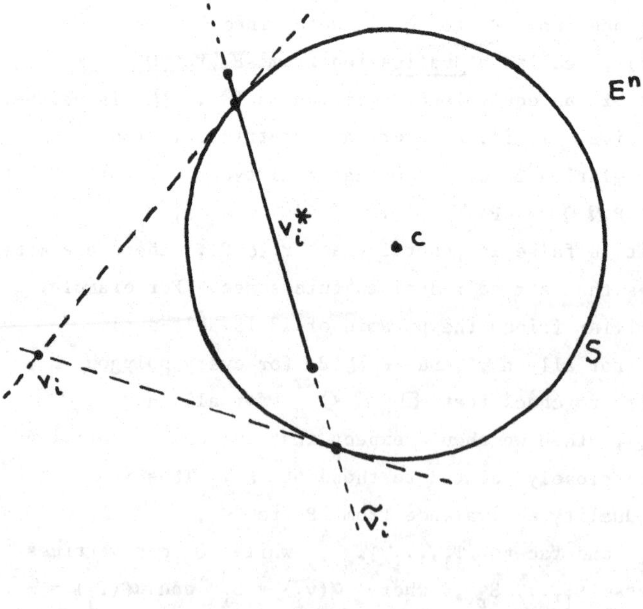

The polytope P_ρ^* is called a polar of P, and $P \wr\wr P_\rho^*$ since there is an anti-isomorphism $\alpha: F(P) \to F(P_\rho^*)$ given by assigning to each j-face T of P with vertices $w_1, \ldots, w_k$ the (n-j-1)-face T^* common to $w_1^*, \ldots, w_k^*$.

The relationship between P and $(P_\rho^*)_\sigma^*$ will be examined more closely in Chapter 3. For the moment we note that $(P_\rho^*)_\sigma^* \approx P$. Also it is an elementary exercise to check that if $P \sim Q$, then for any $\rho > 0$ and any $\sigma > 0$, $P_\rho^* \sim Q_\sigma^*$.

We denote P_ρ^* by P^* in case $\rho = \text{rad } P$ (§3.3). This convention allows us to define a map $\pi: \mathcal{P} \to \mathcal{P}$ by $\pi(P) = P^*$, and the above remarks show that π induces a map $\tilde{\pi}: \mathfrak{S} \to \mathfrak{S}$ by $\tilde{\pi}(\$P) = \$P^*$, where $\$P$ denotes the similarity class of P, as in §1.9. Both π and $\tilde{\pi}$ are bijections. It is important and disenchanting to realise, however, that neither π nor $\tilde{\pi}$ is continuous, as the following example shows.

Let $\varepsilon > 0$ and let P_ε be the polyhedron in E^3 having vertices $(\pm 1, \pm 1, \pm\varepsilon)$. Then P_ε is a cuboid, that is to say $P_\varepsilon \approx \square_3$. It follows that $P_\varepsilon^* \approx \lozenge_3$ is an octahedron. Now $P = \lim_{\varepsilon \to 0} P_\varepsilon$ is a square Q whose polar Q^* is therefore a square also. But $\lim_{\varepsilon \to 0} P_\varepsilon^*$ is not even a polyhedron, but an infinite cylinder with square section. Thus $\lim_{\varepsilon \to 0} \pi(P_\varepsilon) \neq \pi\left(\lim_{\varepsilon \to 0} P_\varepsilon\right)$. If we consider similarity classes instead, then we find that $\lim_{\varepsilon \to 0} \$P_\varepsilon^*$ is the class $*$ of line-segments or 1-polytopes. But again $\lim_{\varepsilon \to 0} \tilde{\pi}(SP\ ) \neq \tilde{\pi}(\$P)$.

We overcome this embarrassment as follows. Let $\mathcal{P}(n)$, $\mathfrak{S}(n)$ denote respectively the spaces of m-polytopes and of similarity classes of m-polytopes, $m \leq n$. Thus $\mathfrak{S}(n) = \mathcal{P}(n)/\sim$, and for all $n \geq 0$, $\mathcal{P}(n-1) \subset \mathcal{P}(n)$ and $\mathfrak{S}(n-1) \subset \mathfrak{S}(n)$. Let $\mathcal{P}^\dagger(n) = \mathcal{P}(n)/\mathcal{P}(n-1)$ and $\mathfrak{S}^\dagger(n) = \mathfrak{S}(n)/\mathfrak{S}(n-1)$. Thus $\mathcal{P}^\dagger(n)$ is obtained from $\mathcal{P}(n)$ by identifying all polytopes of dimension $<n$ to a single point. With these definitions, it follows that π and $\tilde{\pi}$ induce homeomorphisms $\pi_n: \mathcal{P}^\dagger(n) \circlearrowleft$ and $\tilde{\pi}_n: \mathfrak{S}^\dagger(n) \circlearrowleft$. We refer to π_n and $\tilde{\pi}_n$ as polarity homeomorphisms.

Exercise: Prove that $\square_n^* \sim \lozenge_n$ and $\Delta_n^* \sim \Delta_n$.

Exercise: Construct a homeomorphism from $\mathcal{P}_n/\sim$ to $\mathfrak{S}(n)$.

The stratification of $\mathcal{P}(n)$ by combinatorial types induces a corresponding stratification of $\mathcal{P}^{\dagger}(n)$. Let us regard the set of strata of $\mathcal{P}^{\dagger}(n)$ as a poset by putting $S < T$ iff S is contained in the frontier of T. Thus there is a group $\mathcal{K}_n$ of all automorphisms of this poset. The map π_n induces an element of order 2 in $\mathcal{K}_n$, so that $\mathcal{K}_n$ contains a cyclic subgroup of order 2. It seems unlikely that $\mathcal{K}_n$ contains any other elements, so we make the following guess.

DUALITY CONJECTURE <u>$\mathcal{K}_n$ is cyclic of order 2.</u>

7. Joins

The construction that we wish to discuss in §8 is a very special case of one that is familiar to topologists in a more general, abstract, form as the <u>join</u> operation. Since we have occasion to mention this in yet another context in Chapter 4, we give the general definition now. Let X and Y be nonempty topological spaces, and let I denote the closed unit interval $[0,1]$. Define an equivalence relation $\div$ on $X \times Y \times I$ by putting $(x,y,t) \div (x',y',t')$ iff (i) $x = x'$, $y = y'$ and $0 < t = t' < 1$, or (ii) $x = x'$ and $t = t' = 1$ or (iii) $y = y'$ and $t = t' = 0$. Then the quotient space $X{*}Y = (X \times Y \times I)/\div$ is called the <u>join</u> of X to Y. Intuitively, $X{*}Y$ is the space formed by 'joining' each $x \in X$ to each $y \in Y$ by means of an arc, and giving the result the 'obvious' topology.

More concretely, let X and Y be nonempty convex subsets of Euclidean space E such that if $J = \text{aff } X$ and $K = \text{aff } Y$, then $\dim(J \uplus K) = \dim J + \dim K + 1$, and $J \cap K = \emptyset$. (Equivalently, J is <u>skew</u> to K.) Then $\text{conv}(X \cup Y)$ is homeomorphic to $X{*}Y$, and we identify these two spaces in this case. If $[x,y,t]$ denotes the class of $(x,y,t) \in X \times Y \times I$, then the identification is given by mapping $[x,y,t]$ to $tx + (1-t)y$.

If we specialise to the case where the subsets X and Y are polytopes P and Q, then we find that $P{*}Q$ is also a polytope, with $\dim(P{*}Q) = \dim P + \dim Q + 1$. If we make the convention that $X{*}\emptyset = \emptyset{*}X = X$, then the faces of $P{*}Q$ are all of the form $S{*}T$, and again $\dim(S{*}T) = \dim S + \dim T + 1$, where $S \in F(P)$ and $T \in F(Q)$. If $P \approx P'$ and $Q \approx Q'$, then (if defined), $P{*}Q \approx P'{*}Q'$.

It follows that $f_k(P{*}Q) = \sum_{i+j+1=k} f_i(P) f_j(Q)$.

Note also that vert(P*Q) = vert P ∪ vert Q .

Exercise: Prove that if $P \approx \Delta_m$ and $Q \approx \Delta_n$, and aff P is skew to aff Q in the above sense, then $P*Q \approx \Delta_{m+n+1}$.

Exercise: Express $\mu(P*Q)$ in terms of $\mu(P)$ and $\mu(Q)$ for all P , Q for which P*Q is defined as above.

8. Cones

Let P be a polytope and let $v \in E \setminus \text{aff } P$. Then $\Delta_v P = P*\{v\}$ is called a <u>cone</u> with <u>apex</u> v and <u>base</u> P . The combinatorial properties of $\Delta_v P$ are easily deduced from the more general properties of joins.

Thus $\dim \Delta_v P = \dim P + 1$, and for all i ,

$$f_i(\Delta_v P) = f_i(P) + f_{i-1}(P) ,$$

and $\mu(\Delta_v P) = f_0(P) + f_{n-1}(P) + \mu(P)$. In the example shown in Figure 7, where P is a pentagon, $f_{n-1}(P) = f_1(P) = f_0(P) = 5$, $\mu(P) = 10$ and $\mu(\Delta_v P) = 20$.

Figure 7. <u>The cone construction</u>

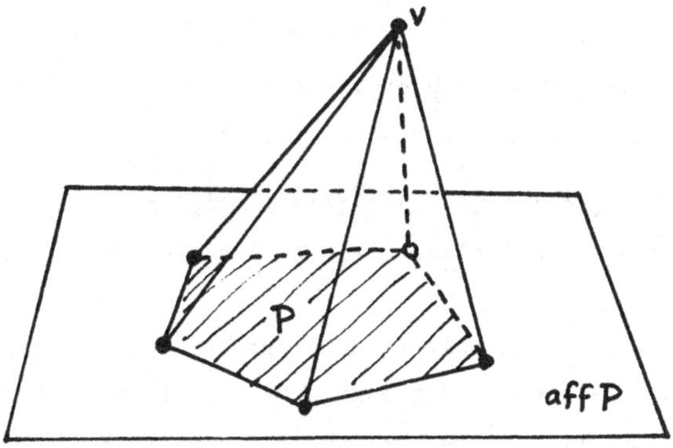

Since $P \approx Q \Rightarrow \Delta_v P \approx \Delta_w Q$ for any appropriately chosen v and w, it follows that $\Delta_v P$ is self-dual whenever P is self-dual.

Exercise: Show that if $P \approx \Delta_n$ and $v \notin \operatorname{aff} P$, then $\Delta_v P \approx \Delta_{n+1}$.

9. Regularity

To study the combinatorial symmetry properties of a polytope P is to study the action of Aut P on the face-lattice $F(P)$. Various concepts of 'regularity' have been developed to isolate those polytopes having particularly high degrees of symmetry. In the next Chapter, we shall be concerned with the metrical symmetry of polytopes, and in anticipation of these later requirements, we discuss regularity in a fairly general setting.

Suppose then that P is an n-polytope, and let G be a subgroup of Aut P. The action of Aut P on $F(P)$ induces an action of Aut P on $F_i(P)$ for each $i = 0, \dots, n-1$, and this restricts to an action of G on $F_i(P)$. We say that P is G-*regular on i-faces* iff this action of G on $F_i(P)$ is transitive.

More generally, for any sequence $J = (j_1, \dots, j_k)$ of integers such that

$$0 \leq j_1 < j_2 < \dots < j_k \leq n-1$$

let $\Phi_J(P) = \Phi_{j_1,\dots,j_k}(P)$ denote the set of J-*flags* $(T_1, \dots, T_k)$, where $T_i \in F_{j_i}(P)$ and $T_1 \triangleleft \dots \triangleleft T_k$. A J-flag for which $J = (0, 1, \dots, n-1)$ is said to be *complete*, and we denote the set of complete flags by $\Phi(P)$. Now G acts on $\Phi_J(P)$ by

$$g\cdot(T_1, \dots, T_k) = (g\cdot T_1, \dots, g\cdot T_k) .$$

We say that P is G-*regular on* J-*flags* iff this action of G on $\Phi_J(P)$ is transitive.

It is of particular interest to consider the case of complete flags. Although it is possible for a proper subgroup of Aut P to act transitively on $\Phi_J(P)$ where $J \neq (0, \dots, n-1)$, the only subgroup of Aut P that can act transitively on the set $\Phi(P)$ of complete flags is Aut P itself. This is because the only combinatorial automorphism of $F(P)$ that fixes a given complete flag of P is the identity. Thus there is at most one element of Aut P mapping any given complete flag to any other. We leave the reader to check these facts as an exercise.

We say then that P is <u>combinatorially regular</u> if $\text{Aut}\,P$ acts transitively (and hence, by the above remarks, simply transitively) on $\Phi(P)$.

A convenient though rather crude indicator of the way in which G acts on $F(P)$ is the G-<u>vector</u>

$$\omega_G(P) = (\omega_0, \ldots, \omega_{n-1}) ,$$

where ω_i is the number of orbits of the action of G on $F_i(P)$. For example, if P is combinatorially regular, then $\omega_G(P) = (1, \ldots, 1)$.

Proposition: <u>Let P be an n-polytope, $n \geq 2$, and for some subgroup G of $\text{Aut}\,P$ let</u>

$$\omega_G(P) = (\omega_0, \ldots, \omega_{n-1}) .$$

<u>Then</u>

$$\omega_0 \leq 2\omega_1 \quad \text{and} \quad \omega_{n-1} \leq 2\omega_{n-2} .$$

Proof: Let $\omega_1 = k$ and let $e_1, \ldots, e_k$ be the edges of P lying in distinct G-orbits in $F_1(P)$. Suppose that e_j has end-points v_j , v_j', $j = 1, \ldots, k$. Let v be any vertex of P . Then v is an end-point of some edge e , say. Now for some $g \in G$ and some $j = 1, \ldots, k$, $e = g \cdot e_j$ and so either $v = g \cdot v_j$ or $v = g \cdot v_j'$. Hence the action of G on $F_0(P)$ has at most $2k$ orbits, that is $\omega_0 \leq 2\omega_1$. That $\omega_{n-1} \leq 2\omega_{n-2}$ follows by duality: we know that if $\omega_G(P^*) = (\omega_0^*, \ldots, \omega_{n-1}^*)$, then $\omega_0^* \leq 2\omega_1^*$. But $\omega_0^* = \omega_{n-1}$ and $\omega_1^* = \omega_{n-2}$.

Corollary: <u>Let P be a polyhedron, and let G be a subgroup of $\text{Aut}\,P$. Suppose that G acts transitively on the set of edges of P . Then</u>

$$\omega_G(P) = (2,1,2) , (2,1,1) , (1,1,2) \text{ or } (1,1,1) .$$

In fact, the first of the four possibilities listed in the above Corollary can be removed, as we now show. The argument given here is due in essence to D.R.J. Chillingworth. Let us say that a subgroup G of $\text{Aut}\,P$ is <u>vertex-transitive</u>, <u>edge-transitive</u> or <u>face-transitive</u> on P according as the polyhedron P is G-regular on 0-faces, 1-faces, or 2-faces of P .

Theorem: Let G be a subgroup of Aut P, where P is a polyhedron, and suppose that G is edge-transitive on P. Then G is either vertex-transitive on P or it is face-transitive on P.

Proof: Let us suppose that $\omega_G(P) = (2,1,2)$, and let the faces of P be labelled T or T' according as they lie in one or other of the orbits of G on $F_2(P)$, and let the vertices be labelled v or v' according to the same criterion. Now each edge e is incident with two faces and has two vertices incident with it. Thus we have the configuration shown in Figure 8. It follows that each vertex has even valency, that is is incident with an even number of edges (and faces), and each face is a polygon having an even number of edges. But this is impossible, since any polyhedron P has either at least one triangular face or at least one vertex of valency 3. For let

$$p = 2f_1(P)/f_0(P) \quad \text{and} \quad q = 2f_1(P)/f_2(P)$$

be the average numbers of edges at each vertex of P and of edges in each face of P. Since, by Euler's Theorem, $f_0(P) + f_2(P) = 2 + f_1(P)$, we find that

$$(p-2)(q-2) < 4 ,$$

Figure 8. Edge-configuration when $\omega_G(P) = (2,1,2)$.

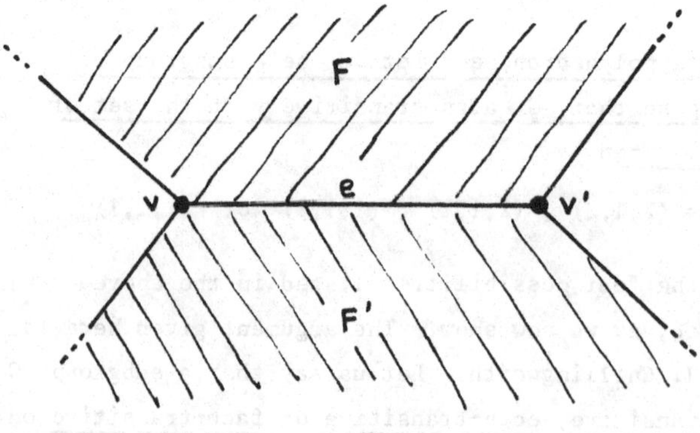

and so either $p < 4$ or $q < 4$. This completes the proof of the theorem.

Exercise: Adapt the above argument to show that if P is a combinatorially regular polyhedron, then P is combinatorially equivalent to one of the five Platonic solids.

The reader will notice that the above theorem does not involve the fact that P is a convex polyhedron. It is a topological theorem about certain kinds of graphs on the 2-dimensional sphere. We are led to ask corresponding questions about analogous graphs Γ on an arbitrary compact connected surface M without boundary, as follows.

Let Γ be a finite graph on such a surface M, having the property that each connected component of $M \setminus \Gamma$ is homeomorphic to the interior of a polygon. Call the closure of each such polygon a face of (M,Γ). We require also that any two faces do not intersect, or have just one vertex in common or have just one edge in common. Let $F(M,\Gamma)$ be the face-lattice defined by analogy with that of a polyhedron, and let G be a subgroup of $\operatorname{Aut} F(M,\Gamma)$. Suppose that G is edge-transitive on (M,Γ). Does it follow that G is either vertex-transitive or fact-transitive? It is known [Jones and Singerman, 1978] that if M is oriented, and G preserves face-orientations, then the question has an affirmative answer. The answer in general is 'no', however. This fact can be deduced from a study of the following ingenious example, due to G.A. Jones.

For a cube K, consider the graph Γ whose vertices are the vertices of K and the mid-points of the edges of K, and whose edges are the pairs (v,v'), where v is the mid-point of some edge e of K, T is a square face of K containing e, and v is a vertex of T that is not an end-point of e. Thus the edges of Γ that lie on a square face of K form a nonconvex octagon as shown in Figure 9.(i). Now Γ can be embedded in a closed orientable surface M of genus 8 with 8 vertices of valency 6, 12 of valency 4, 6 octagonal faces, 8 hexagonal faces and 48 edges. Each hexagon face is bounded by a set of six edges of Γ forming a nonconvex hexagon as shown in Figure 9 (ii). The graph Γ does not quite meet our requirements. For although, by its construction, there is a subgroup G of $\operatorname{Aut}(M,\Gamma)$ that acts transitively on the edges of Γ, and G cannot act transitively on vertices or on faces, each octagonal face has two edges in common with each of its hexagonal neighbours. This defect may be overcome by taking a suitable covering space $\tilde{M}$ of M and showing that the covering graph $\tilde{\Gamma}$ on $\tilde{M}$ has the properties we require.

Examples of combinatorially regular and (metrically) regular polytopes are not difficult to find. Every n-polytope for $n \leqslant 2$ is combinatorially regular, and a polyhedron P is combinatorially regular iff it is combinatorially equivalent to one of the five Platonic solids. In fact, the analogous statement is true in any dimension: any polytope P is combinatorially regular iff it is metrically regular [McMullen, 1967]. Thus there are six combinatorial types of combinatorially regular 4-polytopes, and 3 combinatorial types of combinatorially regular n-polytopes for each $n \geqslant 5$. (Metrical regularity is defined on p. 44.)

Figure 9. Jones's map

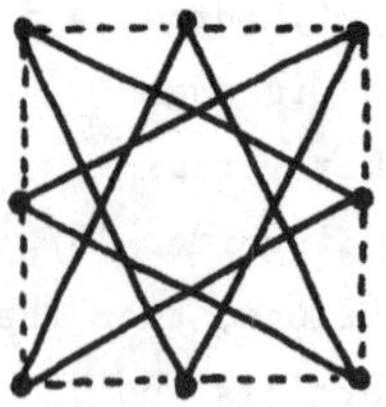

(i) A Jones octagon

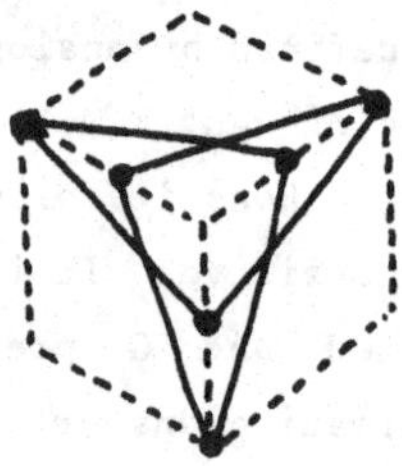

(ii) A Jones hexagon

3. SYMMETRY EQUIVALENCE

The metrical symmetry properties of a polytope P are embodied in the action of its symmetry group $\Gamma(P)$ on the face-lattice $\mathcal{F}(P)$. The associated equivalence relation $\simeq$, which we call symmetry equivalence, stratifies the space $\mathfrak{S}$ of similarity classes topologically. Thus each symmetry equivalence class or symmetry type is a topological manifold whose frontier in $\mathfrak{S}$ is made up of finitely many symmetry types of lower dimension. The topological invariants of each symmetry type are symmetry invariants of the corresponding polytopes. A simple example is the dimension, which we call the deficiency def P of any representative polytope P . Those polytopes P for which def P = 0 are said to be perfect. We show that each (metrically) regular polytope is perfect. However, there are many perfect nonregular polytopes.

1. Transformation groups

The language of group actions has been used extensively in the discussion so far, but little use has been made of any of the associated theory of transformation groups. It is now time to explain some aspects of this theory in order to explore the symmetry type structure of $\mathfrak{S}$. The reader is referred to Bredon [1972] and Borel [1960] for full details of theorems that can only be stated here. We confine attention to actions of compact Lie groups on manifolds without boundary. It is not necessary to include differentiability assumptions.

Suppose that H is a closed subgroup of a compact Lie group G , and let Y = G/H be the left coset space, with projection $\theta : G \to Y$. Then Y is a compact manifold of dimension dim G − dim H , on which G acts transitively by $g \cdot g'H = (gg')H$. For each $y \in Y$, the subgroup $G_y = \{g \in G : g \cdot y = y\}$ of G that fixes y is called the isotropy group at Y . In particular, if $y = \theta(H) = H$, then $G_y = H$. More generally, for any $z \in Y$, G_z is conjugate to H in G . In fact, if $y, z \in Y$, then $z = g \cdot y$ for some $g \in G$, and $G_z = g^{-1} G_y g$.

Conversely, if M is any manifold on which such a group G acts transitively, then M can be identified with G/H , where $H = G_x$ is the isotropy group at some $x \in M$. A diffeomorphism $\mu : M \to G/H$ is given by $\mu(y) = gH$, where $g \in G$ is such that $g \cdot x = y$.

A manifold M together with a transitive action by a (in our case, compact) Lie group G on M , is called a homogeneous space.

More generally, a <u>compact transformation group</u> is a pair (M,G), where M is a manifold and G a compact Lie group acting on M. Thus a homogeneous space is a transformation group of the form $(G/H,G)$.

For any compact transformation group (M,G), and for each $x \in M$, the orbit $G \cdot x$ through x is a compact homogeneous space embedded in M. The concept of isotropy group G_x is then defined as above. Of course it may happen that for some $z \in G \cdot x$, the isotropy groups G_z and G_x are conjugate in G. It makes sense therefore, to introduce the concept of 'orbit type', either in M or in the orbit-space M/G, as follows.

Let H be a closed subgroup of G. Then the <u>orbit type</u> $\omega(H)$ of H is the subset of M consisting of all points $x \in M$ such that G_x is conjugate in G to H. It follows that each orbit type is a union of G-orbits, and so it makes even better sense to consider the set $\omega_*(H) = \omega(H)/G$. The set $\omega_*(H)$ is also called the <u>orbit type</u> of H, and we shall favour this usage.

We have now partitioned M/G into disjoint orbit types, on each which the conjugacy class of G_x is constant. Of course some closed subgroups of G may yield empty orbit types, and we can ignore these.

It may be shown [Borel,1960] that (M,G) has locally finite orbit structure in the sense that each point of M/G has a neighbourhood having nonempty intersections with only finitely many orbit types.

2. Slices

The concept of slice is extremely useful in transformation group theory generally, and is particularly so for the application that we have in mind to the theory of symmetry types. First we explain the auxiliary notion of 'local section'.

Let G be a compact Lie group as before, and let H be a closed subgroup of G, with $Y = G/H$. A <u>local section</u> in Y is an immersion $\sigma : U \to G$, where U is an open neighbourhood of $y = \theta(H)$ in Y, $\sigma(y)$ is the identity element e of G, and $\theta \circ \sigma$ is the identity map 1_Y of Y to itself. The existence of local sections in this sense is a consequence of the fact that θ is a fibre projection whose fibres are the cosets gH. Figure 1 illustrates this idea schematically.

Suppose now that (M,G) is a compact transformation group, and let H be a closed subgroup of G . An H-<u>slice</u> in (M,G) is a subset S of M such that:

(i) S is setwise invariant under H ;

(ii) for all $g \in G \setminus H$, $(g \cdot S) \cap S = \emptyset$;

(iii) for any local section $\sigma: U \to G$ in G/H , the map $\Sigma: U \times S \to M$ given by $\Sigma(u,s) = \sigma(u) \cdot s$ is a homeomorphism from $U \times S$ onto an open subset of M .

A <u>slice</u> at x is a G_x-slice such that $x \in S$.

Example: Let $M = E^2$ and $G = SO(2) = \{A \in O(2) : \det A = 1\}$.
The elements of SO(2) are the matrices

$$A = \begin{bmatrix} \cos\theta & -\sin\theta \\ \sin\theta & \cos\theta \end{bmatrix}, \quad \theta \in R ,$$

and SO(2) is diffeomorphic to the circle S^1 . Consider the transformation group $(E^2, SO(2))$ in which SO(2) acts on E^2 in the standard way, so that $A \cdot x = y$, where

$$y_1 = x_1 \cos\theta - x_2 \sin\theta ,$$
$$y_2 = x_1 \sin\theta + x_2 \cos\theta .$$

Figure 1. <u>Local section</u>

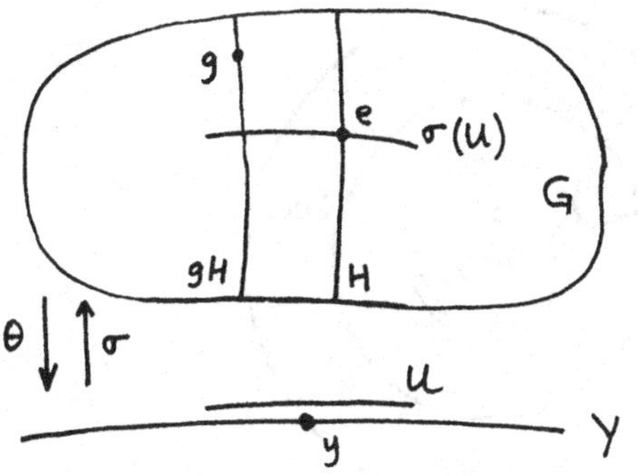

Then the orbits of this action are concentric circles with centre O , except for the orbit through O itself, which is the singleton $\{O\}$. A slice at O is an open neighbourhood of O in E^2 , while a slice at $x \neq O$ is an open straight arc in E^2 containing x and lying in the ray through x from O . In the first case, the image of Σ coincides with the slice, while in the second it is the 'curved rectangle' indicated in Figure 2.

The above is a very simple case of a Euclidean transformation group (M,G) , where $M = E^n$ and G is a subgroup of $O(n)$, the action being obtained by restriction from the standard action of $O(n)$ itself. It may be shown that any transformation group (M,G) of the kind that we have been considering has a locally finite orbit structure and admits an equivariant embedding in $(E^n,O(n))$, for some n . This means that there is a topological embedding $f:M \to E^n$ and a group monomorphism $\phi:G \to O(n)$ such that, for all $x \in M$ and all $g \in G$, $f(g\cdot x) = \phi(g)\cdot f(x)$. Among the consequences of this fact are that (M,G) admits a slice S at each $x \in M$. Moreover, S can be so chosen that, in a suitable coordinate system for M in a neighbourhood of x , G_x acts linearly and S is an open disc in some invariant linear subspace L of the space of coordinates. Hence the fixed point set of G_x on S is the intersection with S of some linear subspace of L .

Figure 2. Slices of $(E^2,SO(2))$.

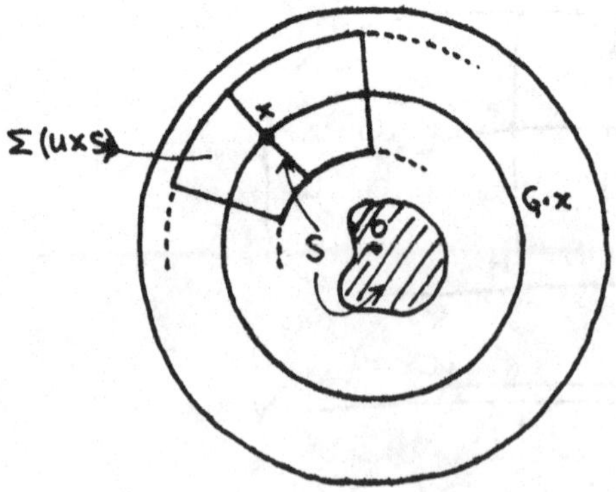

There is a considerable body of theory underlying these results, due to Montgomery, Zippin, Koszul, Gleason, Conner, Floyd, Mostow, Palais and others. For more details and references, see Bredon [1972], Borel [1960] and Montgomery-Zippin [1955].

3. Normal polytopes

To apply the theory of orbit types to the study of symmetry in polytopes, we have to produce a suitable compact transformation group (M,G) . The spaces $\mathcal{P}$, $\mathfrak{S}$, $\mathcal{P}(n)$ and $\mathfrak{S}(n)$ are not manifolds, and the group Sim , though it has many of the features of a Lie group, is neither compact nor finite-dimensional. We have shown, however, that the principal sub-type $\mathbb{C}_*P$ of any n-polytope is a manifold, and the action of Sim(n) on $\mathcal{P}(n)$ restricts to this combinatorial type. Now Sim(n) is certainly a Lie group, but it is not compact. We now squirm out of this difficulty, replacing Sim(n) by the orthogonal group $O(n)$, and replacing $\mathbb{C}_*P$ by a submanifold of 'normal' polytopes. These are obtained by factoring out the effects of dilations and translations, to whose presence in Sim(n) the noncompactness of this group is due.

Translations in E form a subgroup T , and dilations a subgroup D of Sim . They generate a subgroup K of Sim which we can represent as a semidirect product $T.D$, in which multiplication is given by

$$(a,s)(b,t) = (sb+a, st) .$$

Thus K acts on E by

$$(a,s)\cdot x = sx + a .$$

Of course T may be identified with E itself as an additive group, and D with the multiplicative group R_* of positive real numbers.

The action of K on E induces an action of K on $\mathcal{P}$. Now (a,s) has no fixed points if $a \neq 0$ and $s = 1$, and fixes only $(1-s)^{-1}a$ for $s \neq 1$, so it follows that the action of K on $\mathcal{P}^+$ is free. Let $\mathcal{N}_* = \mathcal{P}^+/K$, and let $\nu_*:\mathcal{P}^+ \to \mathcal{N}_*$ be the projection. Then ν_* has a section $\sigma:\mathcal{N}_* \to \mathcal{P}^+$, whose image we call the space $\mathcal{N}$ of 'normal polytopes'. This is constructed as follows.

For any n-polytope P with $n \geq 0$ and having vertices $v_1, \ldots, v_r$, define the <u>centroid</u> cent P of P and the <u>radius</u> rad P of P by

$$\text{cent } P = \frac{1}{r} \sum_{i=1}^{r} v_i ,$$

$$\text{rad } P = \frac{1}{r} \sqrt{\left\{ \sum_{i=1}^{r} \| v_i - \text{cent } P \|^2 \right\}} .$$

Then rad $P = 0$ if $n = 0$ and rad $P > 0$ otherwise. We say that P is normal iff cent $P = 0$ and rad $P = 1$. Thus the set $\mathcal{N}$ of all normal polytopes is a subset of the space $\mathcal{P}^+$ of n-polytopes with $n \geq 1$.

We can define a retraction $\nu: \mathcal{P} \longrightarrow \mathcal{N}$ by $\nu(P) = (a,s)\cdot P$, where $a = -$ cent P and $s^{-1} =$ rad P, and this construction defines the section σ. Thus $\sigma(\nu_*(P)) = \nu(P)$ and $\nu(\nu(P)) = \nu(P)$, for all $P \in \mathcal{P}^+$.

For any $f \in$ Sim, rad$(f\cdot P) = |f|$rad P. Thus $\mathcal{N}$ is not invariant under the action of Sim. However, $\mathcal{N}$ is invariant under O, and it follows at once from these remarks that there is a homeomorphism $h: \mathfrak{S}^+ \to \mathcal{N}/O$ given by $h(\text{Sim}\cdot P) = O\cdot\nu(P)$.

Everything that we have said in the above can be repeated for polytopes in E^n rather than in E. Let $\mathcal{N}(n) = \mathcal{N} \cap \mathcal{P}(n)$. Then ν restricts to $\nu_n: \mathcal{P}(n) \to \mathcal{N}(n)$, and $O(n)$ acts on $\mathcal{N}(n)$ in such a way that h restricts to a homeomorphism $h_n: \mathfrak{S}(n) \to \mathcal{N}(n)/O(n)$ with $h_n(\text{Sim}(n)\cdot P) = O(n)\cdot\nu_n(P)$.

For any polytope P in E^n, ${}_P(n)$ is the submanifold of C_*P defined by the equations cent $P = 0$ and rad $P = 1$. of codimension $n + 1$, as we can check by the method of §2.3. In this way we find that ${}_P(n)$ is a differentiable manifold of dimension $n(f_o(P)+f_{n-1}(P)) - \mu(P) - n - 1$, on which $O(n)$ acts differentiably.

4. Symmetry equivalence

Let P be an n-polytope and let $X =$ aff P. An isometry $g \in$ Iso is said to be a symmetry of P iff $g\cdot P = P$ and $g\cdot Y = Y$ for every affine n-plane Y such that $X \| Y$. It follows that the set $\Gamma(P)$ of all symmetries of P is a finite subgroup of Iso, acting effectively on P. Consequently, this action induces an effective action of $\Gamma(P)$ on $F(P)$. Thus there is an embedding $\alpha_P: \Gamma(P) \to$ Aut P. To distinguish between $\Gamma(P)$ as a group of isometries of P and its image under α_P as a group of combinatorial automorphisms of P, we put $\alpha_P(\Gamma(P)) = \Gamma_*(P)$.

If $P \sim Q$, then $\Gamma(P)$ is conjugate in Sim (in fact in Iso) to $\Gamma(Q)$. Thus the (conjugacy class of) $\Gamma(P)$ is a similarity invariant.

Example: Let P be the n-polytope in E^n having 2^n vertices $(\pm a_1, \ldots, \pm a_n)$. If $0 < a_1 < \ldots < a_n$, then $\Gamma(P)$ is a group of order 2^n, isomorphic to $C_2 \times \ldots \times C_2$ (n factors), acting simply transitively on vert P.

Suppose now that P and Q are polytopes. Then P is said to be <u>symmetry equivalent</u> to Q, written $P \simeq Q$, iff the action of $\Gamma(P)$ on $F(P)$ is equivalent to that of $\Gamma(Q)$ on $F(Q)$ in the following sense. There is an isometry $f \in \mathrm{Iso}$ and a lattice isomorphism $\lambda : F(P) \to F(Q)$ such that, for all $g \in \Gamma(P)$ and all $T \in F(P)$,

$$\lambda(g \cdot T) = (f^{-1} \circ g \circ f) \cdot \lambda(T) .$$

Thus we require that the diagram

$$\begin{array}{ccc} G(P) \times F(P) & \longrightarrow & F(P) \\ f_* \times \lambda \downarrow & & \downarrow \lambda \\ G(Q) \times F(Q) & \longrightarrow & F(Q) \end{array}$$

commute, where $f_* : G(P) \to G(Q)$ is the isomorphism given by $f_*(g) = f^{-1} \circ g \circ f$, and the horizontal arrows denote the group actions.

The definition shows directly that

$$P \sim Q \Rightarrow P \simeq Q \Rightarrow P \approx Q .$$

None of these implications is reversible, however, as the quadrilateral examples illustrated in Figure 3 demonstrate.

Figure 3. <u>Similarity, symmetry equivalence and combinatorial equivalence</u>

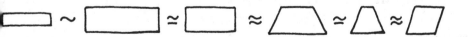

Since $\simeq$ is coarser than $\sim$, we may regard symmetry equivalence as a relation on the space $\mathfrak{S}$ of similarity classes of polytopes, and we shall emphasise this point of view. Our aim is to describe the partition of $\mathfrak{S}$ into symmetry equivalence classes or symmetry types. We denote the symmetry type of $P \in \mathcal{P}$ by $⋇P \subset \mathfrak{S}$.

5. Symmetry types as orbit types

We have seen in §3 that there is a homeomorphism $h_n : \mathfrak{S}(n) \to \mathcal{N}(n)/O(n)$, and that the set $\mathcal{N}_P(n)$ of normal n-polytopes combinatorially equivalent to P is a manifold. Hence the orbit space $\mathcal{N}_P(n)/O(n)$ embeds in $\mathfrak{S}(n)$ under h_n^{-1} with image $\mathbb{C}_*P/\sim$.

The key observation that we make is that for each $Q \in \mathcal{N}_P(n)$, the isotropy group $O(n)_Q$ at Q is just the symmetry group $\Gamma(Q)$. Hence the symmetry types in $\mathbb{C}_*P/\sim$ are the images under h_n^{-1} of the orbit types of $(\mathcal{N}_P(n), O(n))$. Because $(\mathcal{N}_P(n), O(n))$ is a differentiable transformation group, a slice S at P can be chosen in such a way that $\Gamma(P)$ acts linearly on S with respect to suitably chosen local coordinates. In such a coordinate system, the fixed point set F also has a linear (affine) structure. (See Palais' article, p.108, in Borel (1960).). We see therefore that S may be represented as an m-disc in R^m for some $m \geq 0$, and F is the intersection with S of some affine plane. But F is just the intersection with S of the orbit type of P. Hence the orbit-type $\omega_*(\Gamma(P))$ to which P belongs is homeomorphic to a neighbourhood of P in F. That is, $\omega_*(\Gamma(P))$ is a manifold. Since $\omega_*(\Gamma(P))$ maps homeomorphically onto $⋇P$ in $\mathfrak{S}(n)$, we conclude that each symmetry type is manifold. A further conclusion, from the fact that the action of $O(n)$ is locally finite is that the frontier of each symmetry type is a union of finitely many symmetry types of lower dimension. In this sense, $\mathfrak{S}(n)$ and hence $\mathfrak{S}$ is topologically stratified by symmetry equivalence. We denote the set of symmetry types by $\mathcal{S}$, and call $(\mathfrak{S}, \mathcal{S})$ the symmetry stratification of $\mathfrak{S}$. The pair $(\mathfrak{S}, \mathcal{S})$ is now our principal object of study.

Example: Let P be a right prism in E^3 with base a regular m-sided polygon B, where $m \geq 3$ and $m \neq 4$. Then $\Gamma(P)$ is isomorphic to $D_m \times C_2$, where D_m is the dihedral group of degree m (and order 2m).

Let the height of P be h and let r be the radius of B in the sense of §3 (so r is the circumradius of B in this case). We may then denote this prism by $P_m(h,r)$, noting that $P_m(h,r) \sim P_n(k,s)$ iff $m = n$ and $h/r = k/s$. The ratio $\rho = h/r$ is therefore a similarity invariant, and we can parametrise the symmetry type $\Pi =$ ※$P_m(h,r)$ by ρ , with $0 < \rho < \infty$. Thus Π is a connected 1-manifold. The frontier of Π in $\mathfrak{S}$ is obtained by allowing h and r to diminish to 0 in turn. Now $\lim_{h \to 0} P_m(h,r)$ is the m-sided regular polygon B , while $\lim_{r \to 0} P_m(h,r)$ is a 1-polytope. The symmetry types of both are 0-dimensional. Figure 4 illustrates these relationships with $m = 5$.

Figure 4. <u>Examples of symmetry types</u>

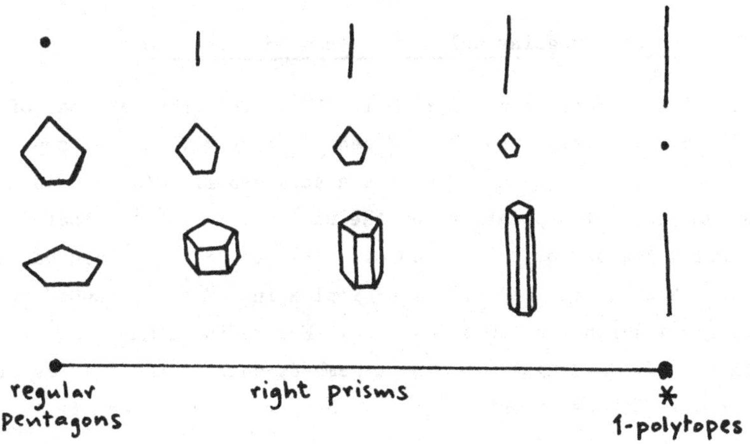

6. Symmetry invariants

The most obvious symmetry invariant of any polytope P is its symmetry group $\Gamma(P)$, or more precisely the conjugacy class of $\Gamma(P)$ in Sim. In practice, it is more interesting to look at $\Gamma_*(P)$ or to represent $\Gamma_*(P)$ as a permutation group on the vertices of P.

Other obvious symmetry invariants are the topological invariants of the symmetry type ⋇P. In principle, we ought to be able to calculate these knowing only the action of $\Gamma(P)$ on $F(P)$. Little progress has been made in this direction. However, it does seem quite clear that a high degree of symmetry for P goes hand-in-hand with a low dimension for ⋇P. More precisely, the dimension for P gives a measure of the lack of transitivity of $\Gamma(P)$ on flags of P.

A polytope P is said to be <u>metrically</u> <u>regular</u> or just <u>regular</u> if the action of $\Gamma(P)$ on $\Phi(P)$ is transitive. Equivalently, P is regular iff it is combinatorially regular and $\Gamma_*(P) = \mathrm{Aut}\ P$. We expect the dimension of the symmetry type of a regular polytope to be very low. In fact, we can easily prove the following.

Theorem: <u>Let P be a regular polytope. Then dim ⋇P = 0.</u>

Proof: Let P be a regular n-polytope in E^n. Then the vertices of P lie on the sphere of radius rad P and centre cent P. We may suppose that P is normal, so that $H = \Gamma(P)$ is a subgroup of $O(n)$, and vert P is the orbit of a point v on the unit sphere S^{n-1}. Let Δ be a fundamental region for the action of $\Gamma(P)$ on S^{n-1}. Then we can take $v \in \Delta$. Now Δ is a spherical n-simplex in S^{n-1}, bounded by portions of great j-spheres, $0 \leqslant j \leqslant n-2$. Now define a map $\phi: S^{n-1} \to \mathfrak{S}$ by $\phi(x) = \$(H\cdot x)$. Then ϕ embeds each of the j-faces T of Δ, $0 \leqslant j \leqslant n$, as a symmetry type. But v is a vertex of Δ, so dim $G\cdot v = 0$.

For a full discussion of these symmetry groups of regular polytopes, see, for example, Coxeter [1948].

We call dim ⋇P = def P the <u>deficiency</u> of P, and say that P is <u>perfect</u> iff def P = 0. The symmetry type of a perfect polytope P, therefore is just the singleton whose only element is \$P. We refer to such symmetry types as <u>nodes</u>. The $\Gamma(P)$-vector $\omega(P)$ of P is called the <u>orbit</u>-<u>vector</u> of P. Thus $\omega(P) = (1,\ldots,1)$ if P is perfect.

While every regular polytope is perfect, not all perfect polytopes are regular. The proof of the above Theorem makes this fact obvious: take a finite subgroup H of $O(n)$ generated by reflexions, and let $P = \text{conv}(H.v)$, where v is a vertex of the fundamental simplex of H in S^{n-1}. Check that $H = \Gamma(P)$. Then P is perfect.

Example: Let H be the octahedral group in $O(3)$. Its fundamental simplex in S^2 is a spherical triangle with three vertices. Their three orbits under H are the vertex sets of a cube, an octahedron and a cuboctahedron. All three are perfect. The cuboctahedron is sketched in Figure 5. This polyhedron is one of the thirteen <u>Archimedean solids</u>. However, not all Archimedean solids are perfect.

There are only nine similarity classes of perfect polyhedra, as we shall see in Chapter 6. Meanwhile, the reader may care to check that if P is an Archimedean solid ($\Gamma(P)$ transitive on vertices, every face a regular polygon), then def P is 0,1 or 2. We shall obtain a simple formula for the deficiency of any polygon in Chapter 5 and discuss a corresponding, even simpler, conjectured formula for the deficiency of a polyhedron in Chapter 6.

Figure 5. <u>A cuboctahedron</u>

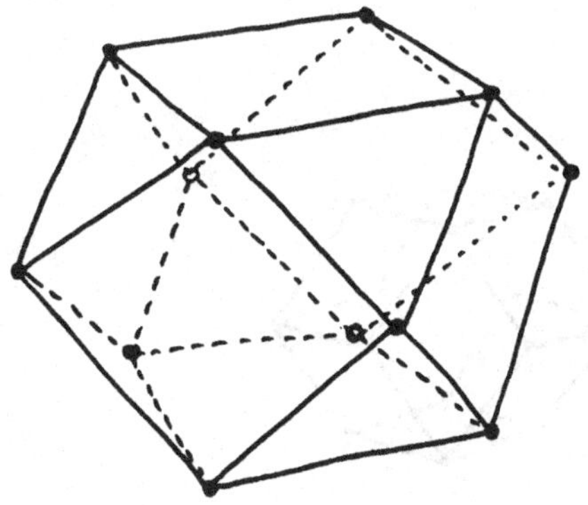

7. Symmetry equivalence and polarity

Once again we take up the theme of §2.6. Let P be a polytope and let $Q = P^*$ be its polar. Then $\Gamma(P) = \Gamma(Q)$. Moreover, if $\alpha: F(P) \to F(Q)$ denotes the anti-isomorphism of face-lattices given by the polar construction of §2.6, then, for all $T \in F(P)$ and all $g \in \Gamma(P)$, $g \cdot \alpha(T) = \alpha(g \cdot T)$. It follows easily that if $P \simeq P'$, then $Q \simeq (P')^*$. For example, the polar of the cuboctahedron of Figure 5 is a rhombic dodecahedron (of the first kind [second edition (1963) of Coxeter, 1948], [Bilinski, 1960]) shown in Figure 6. This too is a perfect polyhedron.

Now $\mathfrak{S}^\dagger(n)$ inherits a symmetry type stratification from that of $\mathfrak{S}(n)$, and we may consider the associated poset whose elements are the strata of $\mathfrak{S}^\dagger(n)$, in which $S < T$ means that S lies in the frontier of T. The subspace $\mathfrak{S}(n-1)$ of $\mathfrak{S}(n)$ is collapsed to a point-stratum of $\mathfrak{S}^\dagger(n)$. The polarity map $\tilde{\pi}_n: \mathfrak{S}^\dagger(n) \circlearrowleft$ induces an automorphism of the poset of order 2. Hence the group $\mathcal{G}_n$ of all automorphisms of the poset of symmetry types in $\mathfrak{S}^\dagger(n)$ contains a subgroup of order 2. As in §2.6, it seems extremely unlikely that any other elements of $\mathcal{G}_n$ exist. We therefore formulate the obvious analogue of the duality conjecture, as follows.

POLARITY CONJECTURE <u>The group $\mathcal{G}_n$ is cyclic of order 2.</u>

Figure 6. <u>A rhombic dodecahedron</u>

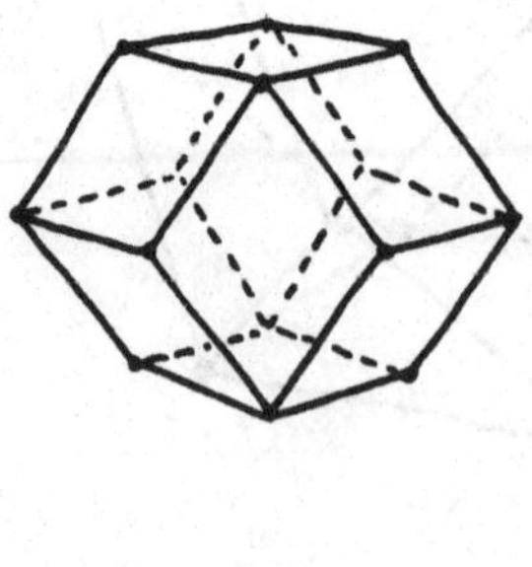

4. PRODUCTS AND SUMS

To some extent the ideas discussed in this chapter are quite old. Schoute [1896] described the construction that we call (following Pólya) the rectangular product. The dual or polar construction here called the rectangular sum seems to have received little or no attention, however. These two operations may be regarded either as operations on combinatorial types or on congruence classes. They do not respect similarity. Nevertheless we can learn a good deal about perfect polytopes from the properties of the rectangular sum and product. In particular, we can show that there are infinitely many similarity classes of perfect 2m-polytopes, for all $m \geqslant 1$.

The key ideas in this chapter were first developed in collaboration with H.R. Morton, but have not been published hitherto.

1. Linear decomposition

Let V be a linear space over a field F , and let X be a subset of V . A linear decomposition of X in V is an ordered set $(X_1, \ldots, X_r)$ of subsets X_i of V such that, if $V_i = \text{aff } X_i$, then:

(i) $X = X_1 + \ldots + X_r$;

(ii) $V = V_1 \oplus \ldots \oplus V_r$;

(iii) for all $i = 1, \ldots, r$,

$\dim V_i \geqslant 1$.

The number r is called the order of such a linear decomposition, and we say that X is linearly decomposable or linearly indecomposable in V according as it does or does not admit a linear decomposition of order $r \geqslant 2$.

A linear decomposition $(X_1, \ldots, X_r)$ of X in V is said to be complete iff , for each $i = 1, \ldots, r$, X_i is linearly indecomposable in V_i . An immediate consequence of these definitions is that if $\dim \text{aff } X \geqslant 1$, so that X contains at least two points, then X admits a complete linear decomposition. The question arises as to whether this decomposition is unique up to ordering of the factors X_i . Obviously the answer is 'no' in general. For example, if $X = V$ and $e_1, \ldots, e_n$ is any basis for V , with $n = \dim V \geqslant 2$, then $(E_1, \ldots, E_n)$ is a complete linear decomposition of V , where $E_i = \langle e_i \rangle$ is the line generated by e_i , $i = 1, \ldots, n$.

A reasonable conjecture, it seems, is the following:

CONJECTURE: Let $(X_1, \ldots, X_r)$ and $(Y_1, \ldots, Y_s)$ be complete linear decompositions of $X \subset V$. Then $r = s$ and for some permutation $\sigma \in S_r$ and some linear automorphism α of V,

$$\alpha(X_i) = Y_{\sigma(i)}, \quad i = 1, \ldots, r.$$

We now show that if X satisfies a very mild condition, then the situation is much less flexible. We borrow and adapt a term from the theory of convex sets: we say that $x \in X$ is an *extreme point* of X if x is not the mid-point of any pair of distinct points x_1, x_2 of X. That is to say, x is an extreme point of X iff there do not exist $x_1, x_2 \in X$ such that $x_1 \neq x_2$ and $2x = x_1 + x_2$. We denote the set of all extreme points of X by ext X.

Theorem: Let $X \subset V$ be such that ext $X \neq \emptyset$, and let $(X_1, \ldots, X_r)$ and $(Y_1, \ldots, Y_s)$ be complete linear decompositions of X. Then $r = s$ and for some $\sigma \in S_r$, $X_i = Y_{\sigma(i)}$, $i = 1, \ldots, r$.

Proof: Let $x \in$ Ext X. Without loss of generality we may suppose that $x = 0$. It follows immediately that 0 is an extreme point of X_i for each $i = 1, \ldots, r$, and $X_i = X \cap V_i$, where $V_i = \text{aff } X_i$ as above. Now let $W = Y_1$ and $Z = Y_2 + \ldots + Y_s$. We have only to show that $W = X_k$ for some $k = 1, \ldots, r$.

Let $X_i \cap W = W_i$, $X_i \cap Z = Z_i$. Then $W_i + Z_i \subset X_i$, since $W_i + Z_i \subset V_i$ and $W_i + Z_i \subset W + Z = X$. Now let $x_i \in X_i$. There exist unique $w \in W$, $z \in Z$ such that $x_i = x + z$. Since $0 \in X$, $W \subset X$ and $Z \subset X$. Hence there are unique points $w_j \in X_j$, $z_j \in X_j$ such that $w = w_1 + \ldots + w_r$ and $z = z_1 + \ldots + z_r$. It follows that $x_i = w + z = (w_1+z_1) + \ldots + (w_r+z_r)$, and so $w_i + z_i = x_i$ and $w_j + z_j = 0$ for $j \neq i$. But 0 is an extreme point of X_j, so $w_j = z_j = 0$ for all $j \neq i$. It follows that $w \in X_i$ and $z \in X_i$, and consequently $w \in W_i$ and $z \in Z_i$. Thus $x_i \in W_i + Z_i$, that is $X_i \quad W_i + Z_i$. We have now proved that $X_i = W_i + Z_i$. But X_i is linearly indecomposable, and therefore either $W_i = 0$ or $Z_i = 0$. Hence $X_i = W$ or $X_i \subset Z$. By applying this argument to $X_1, \ldots, X_r$, we find k such that $X_k = Y$.

Now repeat the whole procedure for $X' = X_1 + \ldots + \hat{X}_k + \ldots + X_r$, and after $r - 2$ such repetitions we obtain the conclusion of the Theorem.

The idea of linear decomposition may be modified by the addition of the extra requirement that the linear subspaces $V_1, \ldots, V_r$ be mutually orthogonal with respect to a given inner product on V. The linear decomposition is then called an orthogonal decomposition of X. The above Theorem may now be rewritten as the following, and may be proved using exactly the same argument.

Theorem: Let $X \subset V$ be such that $\text{ext}\, X \neq \emptyset$, and let $(X_1, \ldots, X_r)$ and $(Y_1, \ldots, Y_s)$ be complete orthogonal decompositions of X. Then $r = s$ and for some $\sigma \in S_r$, $X_i = Y_{\sigma(i)}$, $i = 1, \ldots, r$.

In view of these results, we say that X is linearly prime or orthogonally prime according as X is linearly indecomposable or orthogonally indecomposable. For the sake of simplicity, and because this is the case of most interest for the metrical theory, we concentrate our attention on orthogonal decomposition. As far as our treatment of the combinatorial theory is concerned, it makes no difference whether we consider linear or orthogonal decomposition. For the metrical theory, however, there is a real distinction. Thus any quadrilateral is combinatorially equivalent to a parallelogram and to a rectangle. But whereas both are linearly decomposable, only the rectangle is orthogonally decomposable.

The hypothesis that $\text{ext}\, X \neq \emptyset$ is satisfied by, for example, compact subsets of V. In particular, for any polytope P, $\text{ext}\, P = \text{vert}\, P$. Hence $\text{ext}\, P \neq \emptyset$ whenever $P \neq \emptyset$. If $(P_1, \ldots, P_r)$ is an orthogonal decomposition of a polytope P, then each P_i is itself a polytope.

2. The rectangular product

The above discussion of linear and orthogonal decomposition leads us naturally to a binary operation on polytopes that leads to a number of interesting combinatorial and metrical propositions. We define this concept as follows, in a form to suit our purposes.

Let $\pi: E \times E \to E$ be the linear isomorphism given by $\pi(x,y) = z$, where, for all $i \geq 1$, $z_{2i-1} = x_i$ and $z_{2i} = y_i$. Then $E_1 = \pi(E\times O)$ and $E_2 = \pi(O\times E)$ are mutually orthogonal complementary linear subspaces of E, so that $E = E_1 \oplus E_2$.

Suppose now that X and Y are subsets of E. Then we define the <u>rectangular product</u> $X \square Y$ of X with Y by

$$X \square Y = \pi(X\times Y) .$$

As we have already observed implicitly in §1 above, $\text{ext}(X\square Y) = (\text{ext } X)\square(\text{ext } Y)$. Likewise, $\text{conv}(X\square Y) = (\text{conv } X)\square(\text{conv } Y)$. It follows that if P and Q are polytopes, then so is $P \square Q$.

It follows immediately from the definition of $\square$ that if $P \equiv P'$ and $Q \equiv Q'$, then $P \square Q \equiv P' \square Q'$, and if $P \approx P'$ and $Q \approx Q'$, then $P \square Q \approx P' \square Q'$. It should be noted, however, that $P \sim P'$ and $Q \sim Q'$ do <u>not</u> imply that $P \square Q \sim P' \square Q'$. For example, let P and Q be line-segments. Then the similarity class of the rectangle $P \square Q$ is determined by the ratio of the lengths of P and Q.

We might hope, therefore, to develop useful combinatorial and congruence theories for $\square$. It is convenient to regard $\square$ as an operator on combinatorial classes, given by

$$\mathbb{C}P \square \mathbb{C}Q = \mathbb{C}(P\square Q) ,$$

and on congruence classes, given by

$$\mathbb{K}P \square \mathbb{K}Q = \mathbb{K}(P\square Q) .$$

In both cases, $\square$ is commutative and associative, while $\emptyset$ and Θ, the classes of the empty polytope and of 0-polytopes respectively, act as a zero and an identity. Thus for any polytope P,

$$\emptyset \square P = \emptyset ,$$

and for any singleton $\{x\}$,

$$\{x\} \square P \approx P .$$

3. Combinatorial structure

We now say that a polytope P is $\Box$-*decomposable* if there are polytopes P_1, P_2 such that $P \approx P_1 \Box P_s$, where $\dim P_i \geq 1$, $i = 1,2$, and that P is $\Box$-*indecomposable* if no such P_1, P_2 exist. A $\Box$-*decomposition* of P is then a sequence $(P_1, \ldots, P_r)$ of polytopes P_i such that $\dim P_i \geq 1$ and $P \approx P_1 \Box \ldots \Box P_r$. Such a $\Box$-decomposition is *complete* iff each P_i is $\Box$-indecomposable.

Obviously we should like to have a unique $\Box$-decomposition theorem for polytopes up to combinatorial equivalence and it is natural to make the following guess.

CONJECTURE: *Let $(P_1, \ldots, P_r)$ and $(Q_1, \ldots, Q_s)$ be complete $\Box$-decompositions of a polytope P. Then $r = s$ and for some $\sigma \in S_r$, $P_i \approx Q_{\sigma(i)}$, $i = 1, \ldots, r$.*

Although we have not succeeded in settling this conjecture one way or the other, we can at least explore the behaviour of various combinatorial invariants with respect to the rectangular product.

4. Lattice products

If P and Q are polytopes, what is the relation between $F(P)$, $F(Q)$ and $F(P \Box Q)$? The answer is not quite as simple as we might hope.

Let us recall some standard operations in the category of sets with base point. The objects in this category are pairs (X,x), where X is a set and $x \in X$, and a morphism from (X,x) to (Y,y) is a map $f:X \to Y$ such that $f(x) = y$. It is customary to denote all base-points by $*$ and omit them from the symbol: thus $(X,*) = (X,x) = X$. With these conventions, we define the Cartesian product $X \times Y$ of X and Y in the usual way, choosing $* = (*,*) \in X \times Y$.

The subset $X \times \{*\} \cup \{*\} \times Y$ of $X \times Y$ is denoted by $X \vee Y$. Its base-point, as befits a subobject, is the same as that of $X \times Y$ itself, namely $* = (*,*)$. The quotient set $X \wedge Y = (X \times Y)/(X \vee Y)$ is obtained from $X \times Y$ by collapsing $X \vee Y$ to a single point $*$. These operations, familiar to topologists, are illustrated in Figure 1 for the case in which X and Y are line-segments and $*$ is an end-point.

Let us now consider these operations in the slightly more special case in which our objects are lattices. For a finite lattice L, we have two obvious choices for base-point: we can choose the maximum element or the minimum element. For reasons that will become clearer shortly, we choose the minimum element as base-point. Later we shall see that in the context of 'rectangular sum', the maximum element is the appropriate choice. From now on, all lattices are finite.

Suppose, then that we denote the partial order in any lattice by $<$ and the minimum element by $*$. Let L and M be lattices. Then the Cartesian product $L \times M$ has a lattice structure given by $(\lambda,\mu) < (\lambda',\mu')$ iff $\lambda < \lambda'$ and $\mu = \mu'$, or $\lambda = \lambda'$ and $\mu < \mu'$, or $\lambda < \lambda'$ and $\mu < \mu'$, and $* = (*,*)$. Likewise, $L \vee M$ is a sublattice of $L \times M$, and there is a lattice structure induced in $L \wedge M$.

5. Face-lattice of rectangular products

We now apply these considerations to help us understand the combinatorial structure of $P \square Q$ in terms of the corresponding structures of the factor polytopes P and Q.

Theorem: <u>For any polytopes P and Q, there is a lattice isomorphism</u>

$$\phi : F(P \square Q) \to F(P) \wedge F(Q) .$$

Proof: We have only to observe that every face of $P \square Q$ is of the form $A \square B$, where $A \in F(P)$ and $B \in F(Q)$, while $\emptyset \square B = A \square \emptyset = \emptyset$, and $\emptyset = *$.

Figure 1. <u>The operators $\vee$ and $\wedge$</u>

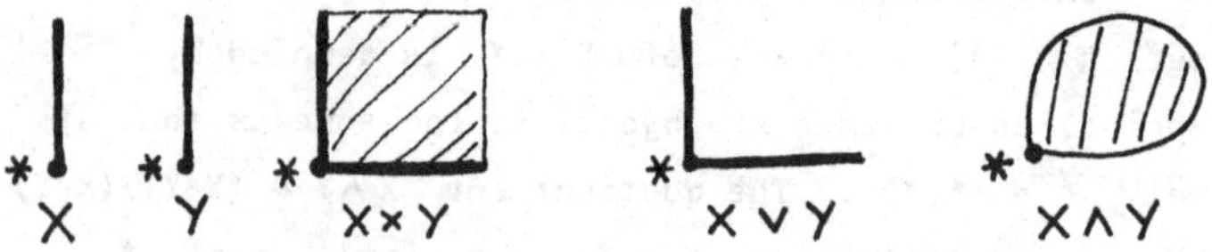

It follows that for all $k \geq 0$, $F_k(P \square Q)$ may be identified with the union of the sets $F_i(P) \times F_j(Q)$, where $i \geq 0$, $j \geq 0$ and $i + j = k$. Hence for all $k \geq 0$, $f_k(P \square Q) = \sum_{\substack{i=0 \\ i+j=k}}^{m} \sum_{j=0}^{n} f_i(P) f_j(Q)$, where $m = \dim P$ and $n = \dim Q$. Note that $\dim(P \square Q) = \dim P + \dim Q$.

As an example, let P be a line-segment and Q a pentagon. Then $P \square Q$ is a prism with pentagonal base. Now $f(P) = (2)$, while $f(Q) = (5,5)$. And $f(P \square Q) = (10,15,7)$, where we note that $10 = 2.5$, $15 = 2$.

From the Theorem we also deduce that

$$\mu(P \square Q) = f_0(P)\mu(Q) + f_0(Q)\mu(P) .$$

5. Combinatorial automorphisms

Let us turn now to questions of combinatorial automorphisms. Let P and Q be polytopes. Then there is an action of $\mathrm{Aut}\, P \times \mathrm{Aut}\, Q$ on $F(P) \times F(Q)$ which induces an action of this group on $F(P) \wedge F(Q)$. In this way we see that $\mathrm{Aut}\, P \times \mathrm{Aut}\, Q$ embeds naturally in $\mathrm{Aut}(P \square Q)$. However, we cannot expect that this embedding should be an isomorphism unless P and Q have no common $\square$-factors. For example, $\mathrm{Aut}(P \square P)$

Figure 2. <u>The rectangular product</u>

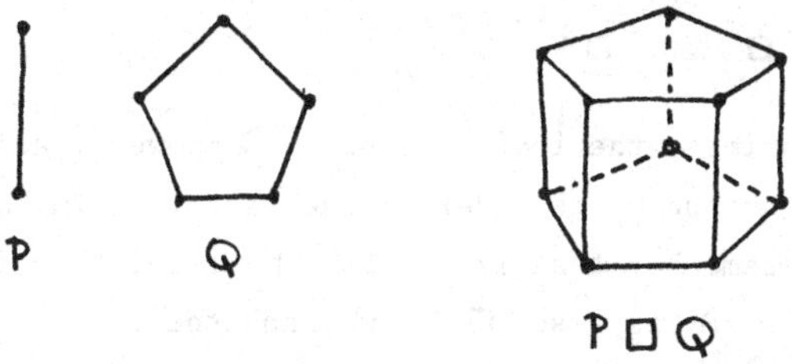

contains the automorphism of $F(P \square P)$ induced by the switch map $(\lambda,\mu) \to (\mu,\lambda)$ on $F(P) \times F(P)$.

More generally, let $(P_1,\ldots,P_r)$ be a complete $\square$-decomposition of P . Then we can collect together those factors that are combinatorially equivalent, to obtain $P \approx Q_1 \square \ldots \square Q_s$, where $Q_i \approx K_i^{r_i} = K_i \square \ldots \square K_i$ (r_i factors), and $r_1 + \ldots + r_s = r$. Now Aut Q_i contains a subgroup isomorphic to the wreath product Aut $K_i \wr S_{r_i}$, and so Aut P contains a subgroup isomorphic to the direct product $\bigtimes_{i=1}^{s}(\text{Aut } K_i \wr S_{r_i})$ of these groups. An obvious guess is that Aut P coincides with this subgroup. (Here S_n denotes the symmetric group of degree n . For a discussion of wreath products, see, for example, Rose [1978].)

Exercise: Let I be any 1-polytope. For any positive integer n , construct a combinatorial equivalence

$$f: \square_n \to \square^n I = I \square \ldots \square I (n \text{ factors}).$$

7. Symmetry groups

The unique linear decomposition Theorem of §1 tells us, among other things, that there is a well-defined notion of primeness for polytopes with respect to the rectangular product operation. Thus we say that a polytope P is $\square$-prime iff it is $\square$-indecomposable, and the linear decomposition Theorem implies that every polytope P is congruent to a polytope

$$P_1^{r_1} \square \ldots \square P_s^{r_s} ,$$

where $P_1,\ldots,P_s$ are $\square$-primes, the indices denote $\square$-powers, and the $\square$-prime expression is unique up to order of the factors. In this context, it is important to remember that if P is $\square$-prime, then for any real number $\lambda > 0$, $Q = \lambda P$ is also $\square$-prime, and their congruence classes $\mathbb{K}P$, $\mathbb{K}Q$ are counted as distinct $\square$-primes. For example, a square $\square_2$ is a $\square$-square $I \square I = \square^2 I$, but a non-square rectangle has $\square$-decomposition $I \square J$, where I , J are distinct $\square$-primes.

It is quite easy to express the symmetry group of a polytope P in terms of its $\square$-prime factors. There are two simple observations to make. Firstly, if P is $\square$-prime and r is a positive integer, then $\Gamma(\square^r P)$ is isomorphic to the wreath product $\Gamma(P) \wr S_r$. Secondly, if P and Q are $\square$-coprime in the sense that, up to congruence, they have no common $\square$-factor, then $\Gamma(P \square Q)$ is isomorphic to $\Gamma(P) \times \Gamma(Q)$. It follows that if P has $\square$-prime power decomposition $P_1^{r_1} \square \ldots \square P_s^{r_s}$, then $\Gamma(P)$ is isomorphic to

$$\Gamma(P_1) \wr S_{r_1} \times \ldots \times \Gamma(P_s) \wr S_{r_s} .$$

Example: $\Gamma(\square_n)$ is isomorphic to $S_n \wr S_2$. It follows that $\Gamma(\square_n)$ has order $2^n n!$. In particular, $\Gamma(\square_2)$ is isomorphic to the dihedral group D_4, of order 8, and $\Gamma(\square_3)$ has order $2^3 3! = 48$.

8. Deficiency and perfection

What is the relationship between def P, def Q and def($P \square Q$)? In the light of our experience with symmetry groups, we should expect that the answer depends on whether or not P and Q are $\square$-coprime. In fact, for any positive real numbers λ and μ, $\lambda P \square \lambda Q \sim P \square Q$, and

$$\lambda P \square \mu Q \simeq P \square Q ,$$

provided that λP and μQ are also $\square$-coprime. It follows immediately that if P and Q are $\square$-coprime, then def($P \square Q$) = def P + def Q + 1, the extra dimension recording the ratio $\lambda : \mu$. Likewise, if P is $\square$-prime, then def $\square^r P$ = r def P. We conclude that if P has $\square$-prime power decomposition $P_1^{r_1} \square \ldots \square P_s^{r_s}$, then

$$\text{def } P = r_1 \text{def } P_1 + \ldots + r_s \text{def } P_s + s - 1 .$$

These simple facts have at least one important consequence. For if P is perfect, it follows that P has $\square$-prime power decomposition of the form $\square^r Q$, where Q is a perfect $\square$-prime. Conversely, if P is perfect, then so is $\square^r P$, for any positive integer r. Since dim $\square^r P$ = r dim P, we conclude that there are infinitely many similarity classes of perfect 2n-polytopes, for any $n \geq 1$, since any regular polygon is perfect.

Any regular polytope is perfect, and any perfect polytope is a $\square$-prime power, as we have seen. Since all the regular polytopes are known, we may check that $\square_n$ is the only regular n-polytope, $n \geq 2$, that is not $\square$-prime.

9. The rectangular sum

The operator $\square$ has a 'dual' which we denote by $\diamond$ and call the <u>rectangular sum</u>. We do not attempt to develop a theory analogous to the linear decomposition theory that underlies the definition of $\square$, but merely regard $\diamond$ as a binary operation on the space of polytopes.

We make use of the join operation $*$ discussed in §2.7 and the map $\pi : E \times E \to E$ introduced in §2. Writing $[x,y,t]$ for the image of $(x,y,t) \in E \times E \times I$ in $E * E$, we define a map $\sigma_{ab} : E * E \to E$, for any $a,b \in E$, by

$$\sigma_{ab}[x,y,t] = \pi((1-t)x + ta\ ,\ ty + (1-t)b) \quad .$$

Then, for any polytopes P and Q, we put

$$P \diamond Q = \sigma_{ab}(P*Q) \ .$$

We find that $P \diamond Q$ is a polytope with $\dim(P \diamond Q) = \dim P + \dim Q$. Whereas the faces of $P \square Q$ are the form $A \square B$, where $A \in F(P)$ and $B \in F(Q)$, the faces of $P \diamond Q$ are not of the form $A \diamond B$, with the exception, of course, of $P \diamond Q$ itself. However, there is an obvious homeomorphism from $A * B$ to a corresponding face of $P \diamond Q$, for each $(A,B) \in F(P) \times F(Q)$. By a slight abuse of notation, we denote this face of $P \diamond Q$ by $A * B$. In fact, there is a lattice-isomorphism from $F(P) \wedge F(Q)$ to $F(P \diamond Q)$, where we choose maximum elements as base-points.

It follows that for all $k \leq m + n$, where $\dim P = m$ and $\dim Q = n$,

$$f_k(P \diamond Q) = \Sigma f_i(P) f_j(Q) \ ,$$

where the sum is taken over i and j such that $i + j + 1 = k$, $i \geq 0$, $j \geq 0$.

In Figure 3, we indicate the form of $P \diamond Q$ in case P is a line-segment and Q is a pentagon. The combinatorial properties of can all be deduced immediately by using the following theorem in conjunction with the properties of $\Box$.

Theorem: <u>Let P, P' , Q and Q' be polytopes such that $P \wr\wr P'$ and $Q \wr\wr Q'$. Then $P \Box Q \wr\wr P' \Diamond Q'$</u> .

Proof: We need only construct an anti-isomorphism $\zeta : F(P \Box Q) \to F(P' \Diamond Q')$. Since $P \wr\wr P'$ and $Q \wr\wr Q'$, there are anti-isomorphisms $\xi : F(P) \to F(P')$ and $\eta : F(Q) \to F(Q')$. It is routine to check that the required map ζ is given by

$$\zeta(A \Box B) = \xi(A) * \eta(B) \quad \text{for} \quad \dim(A \Box B) > -1 .$$

The metrical properties of $\Diamond$ can be deduced from the following two observations. Firstly, if P and Q are $\Box$-coprime, then P^* and Q^* are $\Diamond$-coprime and $(P \Box Q)^* \simeq P^* \Diamond Q^*$. Secondly, $(\Box^r P)^* \sim \Diamond^r P^*$, for any $\Box$-prime P .

Example: $(\Box_n)^* = (\Box^n I)^* \sim \Diamond_n \sim \Diamond^n I$, where I is any 1-polytope.

Figure 3. <u>The rectangular sum</u>

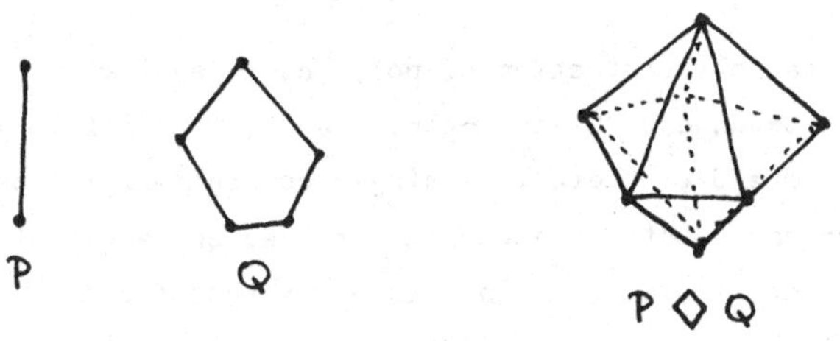

5. POLYGONS

Polygons are extremely simple objects and they have been studied from various points of view throughout the history of Mathematics. It may seem unlikely, therefore, that there is anything both new and interesting to say about them. I hope to convince the reader, however, that this is a false impression. Even in the case of quadrilaterals, the symmetry classification reveals some surprisingly fresh and appealing phenomena. In fact, I have not been able to find any previous classification of quadrilaterals by symmetry, still less of polygons in general, that is much more than a catalogue of 'kinds' of quadrilaterals, such as we find in in Book I of Euclid's Elements. In particular, the important symmetry type of 'kites' or 'deltoids' seems to have been overlooked completely in centuries past: the earliest explicit reference to this shape of quadrilateral appears to occur in the latter half of the nineteenth century.

1. Combinatorial structure

Since polygons are the same things as 2-polytopes, the facets of a polygon P are its edges and the $(n-2)$-faces are its vertices, where $n = 2 = \dim P$. It follows that every vertex belongs to exactly two edges, and of course each edge contains just two vertices. The face-vector of any polygon P is therefore of the form $f(P) = (f_0(P),,f_1(P)) = (m,m)$ for some integer $m \geq 3$. We then refer to P as an m-gon. Thus 3-gons are triangles, 4-gons are quadrilaterals, 5-gons are pentagons and so on.

The combinatorial classification of polygons is easily described: if P is an m-gon and Q an n-gon, then $P \approx Q$ iff $m = n$. Hence, for each integer $m \geq 3$, there is a single combinatorial type Π_m consisting of all m-gons. Let us label the vertices and edges of $P \in \Pi_m$ as $v_1, \ldots, v_m$ and $e_1, \ldots, e_m$ in such a way that for all i (reducing suffices modulo m), $v_i \lhd e_i$ and $v_i < e_{i+1}$, as indicated in Figure 1.

With this labelling, the face-lattice $F(P)$ has the structure indicated in Figure 2 and the incidence matrix is

$$M(P) = \begin{bmatrix} 1 & 0 & 0 & 0 & \dots & & 0 & 1 \\ 0 & 1 & 1 & 0 & \dots & & 0 & 0 \\ 0 & 0 & 1 & 1 & \dots & & 0 & 0 \\ & & & & & & & \\ . & . & . & . & . & . & . & . \\ 0 & 0 & 0 & 0 & \dots & 1 & 1 & 0 \\ 0 & 0 & 0 & 0 & \dots & 0 & 1 & 1 \end{bmatrix} .$$

:he multiplicity of P is therefore $\mu(P) = 2m$, and the dimension of :he principal sub-type $\mathbb{C}_* P$ is $2m$, by the Theorem of §2.3. Of :ourse, since P is both simple and simplicial, we can observe directly :hat $\dim \mathbb{C}_* P = 2m$: for example, each of the m vertices of P has :wo degrees of freedom in aff P. It is also evident, as we remarked .n §2.6, that $P \wr\wr P$. A glance at Figure 2 shows that a suitable anti-isomorphism $\alpha : F(P) \circlearrowleft$ is given by $\alpha(v_i) = e_i$, $\alpha(e_i) = v_{i+1}$.

The group Aut P is isomorphic to the dihedral group D_m of legree m. To obtain an explicit isomorphism, recall that D_m has the presentation

$$D_m = \langle a,b : a^m = b^2 = (ab)^2 = 1 \rangle$$

and consider the elements g and h of Aut P given by $g(v_i) = v_{i+1}$, $h(v_i) = v_{m-i+1}$, $i = 1, \dots$, where suffices are reduced modulo m.

Figure 1. Labelling a polygon

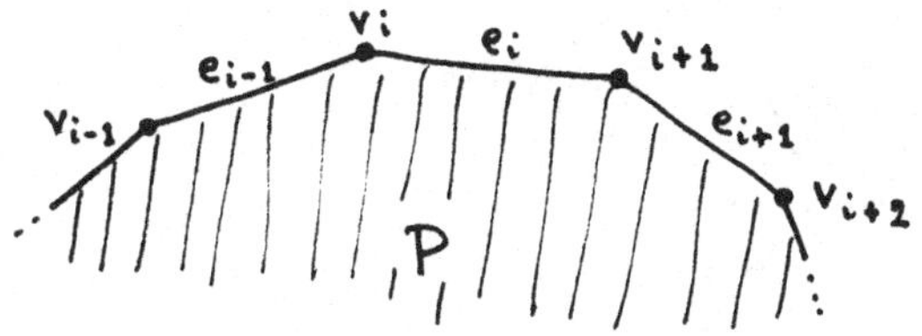

Then g and h generate Aut P , and there is an isomorphism $f:\mathrm{Aut}\,P \to D_m$ for which $f(g) = a$ and $f(h) = b$.

Exercise: Prove that every polygon is combinatorially regular.

2. Possible symmetry groups

Let P be an m-gon. Since $\mathrm{Aut}\,P \approx D_m$, $\Gamma(P)$ is isomorphic to a subgroup of D_m . Now the subgroups of D_m are

$$Z_{m,r} = \langle a^r \rangle , \quad D_{m,r} = \langle a^r, b \rangle \text{ and } D'_{m,r} = \langle a^r, ab \rangle ,$$

where $r|m$. The group $Z_{m,r}$ is isomorphic to the cyclic group C_s of order s , where $rs = m$, while $D_{m,r} \cong D'_{m,r} \cong D_s$. Thus $Z_{m,r}$ is of order s , and both $D_{m,r}$ and $D'_{m,r}$ are of order 2s . If m is even, then $D_{m,r}$ is not conjugate to $D'_{m,r}$ in D_m except when $r = 1$. For m odd, on the other hand, $D_{m,r}$ is conjugate to $D'_{m,r}$.

It follows at once from these remarks that the number γ_m of distinct conjugacy classes of subgroups of D_m is given by

$$\gamma_m = 3\phi - 1 \quad \text{for } m \text{ even,}$$

and

$$\gamma_m = 2\phi \quad \text{for } m \text{ odd,}$$

Figure 2. <u>The face-lattice of a polygon</u>

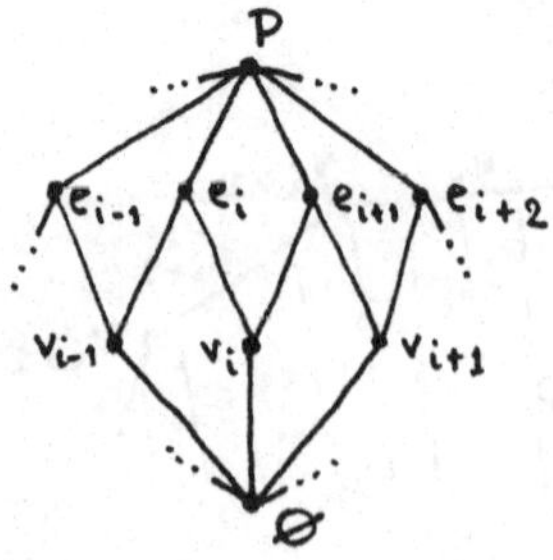

where ϕ is the number of divisors of m. Suppose, then that m has prime power decomposition $p_1^{m_1} \ldots p_k^{m_k}$, where $p_1 < \ldots < p_k$ are prime numbers. Then

$$\phi = (m_1+1) \ldots . (m_k+1) .$$

To illustrate these facts, the first few values of γ_m and of ϕ are tabulated below.

m	3	4	5	6	7	8	9
ϕ	2	3	2	4	2	4	3
γ_m	4	8	4	11	4	11	6

3. Symmetry types

In discussing polygons and their symmetry, we have emphasised the relation between the symmetry group $\Gamma(P)$ of a polygon and the combinatorial automorphism group Aut P rather than the fact that $\Gamma(P)$ is a group of isometries of E, conjugate in Iso to a finite subgroup of $O(2)$. The two approaches may be brought together easily by embedding D_m in $O(2)$, sending $a,b \in D_m$ to $A_m, B \in O(2)$ respectively, where

$$A_m = \begin{bmatrix} c & -s \\ s & c \end{bmatrix} , \quad B = \begin{bmatrix} 1 & 0 \\ 0 & -1 \end{bmatrix} ,$$

$c = \cos\theta$, $s = \sin\theta$ and $m\theta = 2\pi$. It is convenient to identify D_m with its image under this embedding. Our main interest, of course, is in the action of $\Gamma(P)$ on $F(P)$. Thus we identify $\Gamma(P)$ as a subgroup of D_m in $O(2)$. Note, however, that in general the action of D_m on $F(P)$ is not induced by this orthogonal representation of D_m: but the orthogonal action of $\Gamma(P)$ on P induces the action of $\Gamma(P)$ on $F(P)$.

Suppose then that P and Q are polygons, with labelling of vertices and edges so chosen that identifications Aut P and Aut Q with D_m are given as described above. If $G = \Gamma(P)$ and $H = \Gamma(Q)$ are conjugate subgroups of D_m, then the actions of G and H on $F(P)$ and $F(Q)$ are equivalent to one another, and so, by definition, $P \simeq Q$.

Hence the number σ_m of symmetry types of m-gons is equal to the number of conjugacy classes of subgroups of D_m that occur as symmetry groups of m-gons. Thus $\sigma_m \leq \gamma_m$. In fact, $\sigma_m = \gamma_m - 1$, since the only conjugacy class that has to be excluded is that of $Z_{m,1} \simeq C_m$: there is no m-gon P for which $\Gamma(P) \simeq C_m$.

Exercise: For each subgroup G of D_m with $G \neq Z_{m,1}$, construct an m-gon P such that $\Gamma(P) = G$. (Some examples are given below.)

We may now list the first few values of σ_m , using the values of γ_m tabulated in §2.

m	3	4	5	6	7	8	9
σ_m	3	7	3	10	3	10	5

The thirty four symmetry types of m-gons that have non-trivial symmetry group, $3 \leq m \leq 9$, are illustrated in Figure 3, the order of the symmetry group being indicated in each case.

4. Deficiency

Let P be an m-gon with orbit vector $\omega(P) = (\omega_0(P), \omega_1(P)) = (\alpha, \beta)$. The set of fixed points of the action of $\Gamma(P)$ on the plane aff P is a nonempty affine subspace of aff P , of dimension ξ, say.

Proposition: $\underline{\text{def } P = \alpha + \beta - \xi - 2}$.

Proof: I have not succeeded in devising a simple proof of this proposition, but must rely on a crude case-by-case argument. Instead of giving full details, I deal with two cases, and leave the others to the interested reader. The method is straightforward and *ad hoc*.

Case (i): $\xi = 2$. In this case, $\Gamma(P)$ is trivial, and $\sigma(P)$ is just the principal orbit type of the action of Sim(2) on $\mathbb{C}_*P$. Since dim Sim(2) = 4 and dim $\mathbb{C}_*P = 2m$, we conclude that def P = 2m − 4 . Since $\alpha = \beta = m$, the proposition holds in this case.

Case (ii) Let $\Gamma(P)$ be cyclic, say $\Gamma(P) = Z_{m,r} \simeq C_s$, where rs = m and s > 1 . Then $\xi = 0$. We may assume that cent P = 0 , and $P \subset E^2$. Thus $\Gamma(P)$ is generated by the rotation A_m^r . We may label the vertices of P in such a way that, identifying E^2 with the complex

Figure 3

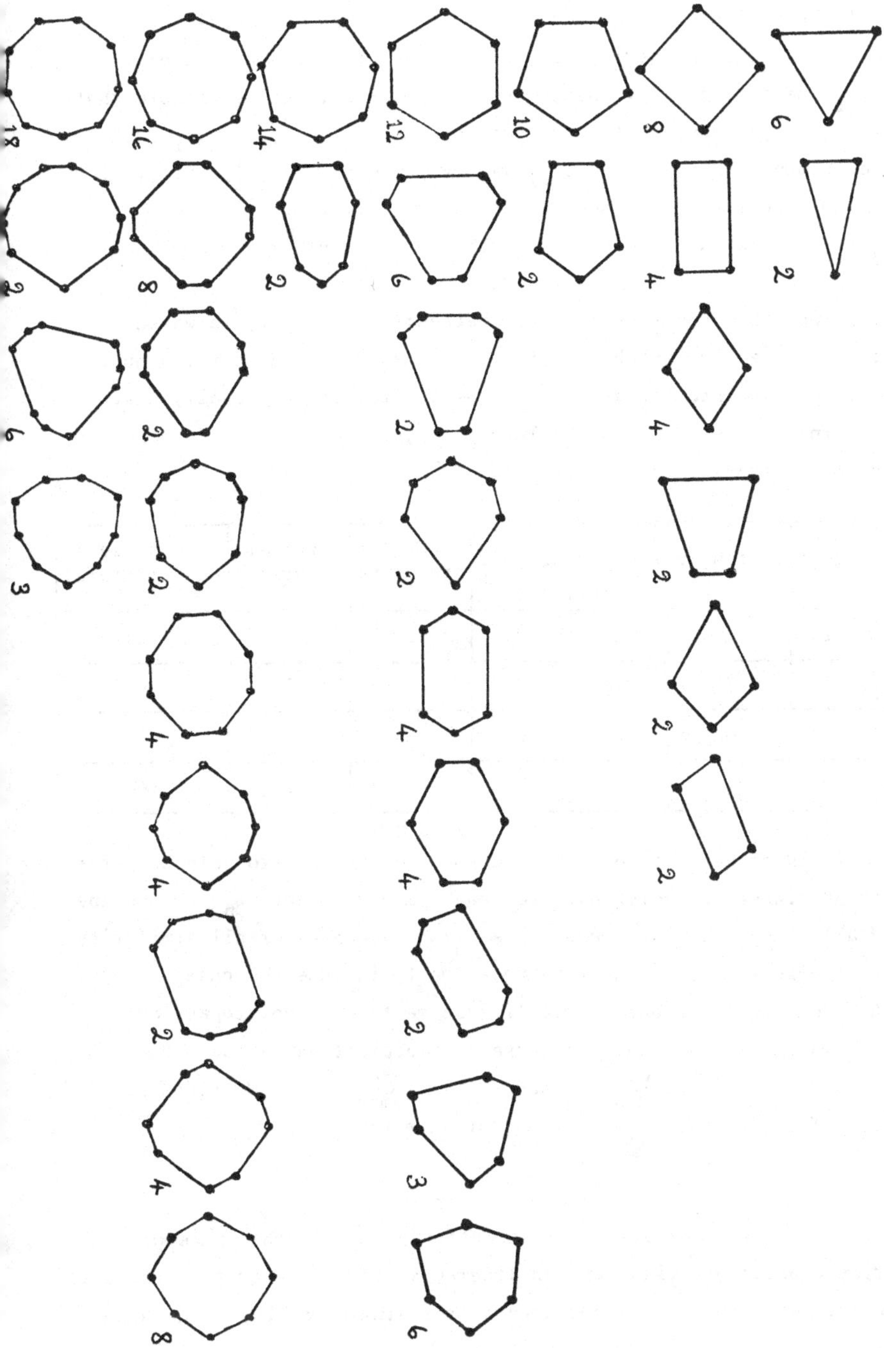

number field $\mathbb{C}$, $v_j = \rho_j e^{i\theta_j}$, $j = 1,\ldots,r$, and $0 \leq \theta_1 < \ldots < \theta_r < 2\pi/s$. Then each point v_j lies in one of the s orbits of the action of $\Gamma(P)$ on $F_0(P)$. Each has two degrees of freedom in E^2 within $\sigma(P)$, and with $\Gamma(P)$ fixed, a choice of $v_1,\ldots,v_r$ determines P . Hence $\operatorname{def} P = 2r - 2$, since we note that if $Q(v_1,\ldots,v_r)$ denotes the polygon determined by $v_1,\ldots,v_r$ and the fixed group $\Gamma(P)$, then $Q(v_1,\ldots,v_r) \sim Q(v_1',\ldots,v_r')$ iff , for some $t > 0$ and some λ , and for all $j = 1,\ldots,r$, $e_j' = te_j$ and $\theta_j' = \theta_j + \lambda$. Now $\alpha = \beta = r$, so $\operatorname{def} P = \alpha + \beta - 2 = \alpha + \beta - \xi - 2$ once again.

Two other cases need to be checked: Case (iii), in which $\xi = 1$; and Case (iv) in which $\Gamma(P)$ is a nontrivial dihedral group. The latter breaks into three subcases, as the reader may readily check.

In illustration, we tabulate α, β, ξ and def P for the seven symmetry types of quadrilaterals P .

P	square	rectangle	rhombus	deltoid	isosceles trapezium	parallelogram	scalene quadrilateral
α	1	1	2	3	2	2	4
β	1	2	1	2	3	2	4
ξ	0	0	0	1	1	0	2
def P	0	1	1	2	2	2	4

It is convenient to introduce some new symbols to help us refer to different classes of polygons. For each $m \geq 3$, let Π_n denote the space of all n-gons as above, and $\Sigma_n = \Pi_n/\sim$ the space of all similarity classes of n-gons. We extend this notation to include the case $n = 2$. Thus Π_2 denotes the space of all line-segments or 1-polytopes, and $\Sigma_2 = \Pi_2/\sim$ denotes the singleton whose only element is the set Π_2 itself. We put $\overline{\Pi}_n = \bigcup_{2\leq m\leq n} \Pi_m$ and $\overline{\Sigma}_n = \bigcup_{2\leq m\leq n} \Sigma_m$. We also denote the space of all polygons by Π and put $\Sigma = \Pi/\sim$.

5. Triangles

We now give a detailed description of $\overline{\Sigma}_3$, the space of similarity classes of triangles and line-segments. We know that there is just one symmetry type of line-segments (the singleton Σ_2) and three symmetry types of triangles, the latter having dimensions 0, 1 and 2 . It is quite easy, in fact, to embed $\bar{\Sigma}_3$ in E^2 , and hence to picture the relationships between these symmetry types, as follows.

Let $T(\alpha,\beta,\gamma)$ denote the similarity class of triangles with (internal) angles α, β, γ , where $\alpha \leqslant \beta \leqslant \gamma$. Since $\alpha + \beta + \gamma = \pi$, the third parameter γ may be discarded, and we can denote the similarity class by $T(\alpha,\beta)$. The parameters α and β can then take any real values, subject to the inequalities

$$0 < \alpha \leqslant \beta , \quad \alpha + 2\beta \leqslant \pi .$$

Thus we may define an embedding $\tau:\Sigma_2 \to E^2$ by $\tau(T(\alpha,\beta)) = (\alpha,\beta)$, whose image in E^2 is the region

$$\Delta = \{x \in E^2 : 0 < x_1 \leqslant x_2 , \quad x_1 + 2x_2 \leqslant \pi\}$$

shown in Figure 4.

The interior of Δ corresponds under τ to the 2-dimensional symmetry type $\mathcal{L}$ of scalene (or 'limping') triangles $T(\alpha,\beta,\gamma)$, $0 < \alpha < \beta < \gamma$. The boundary component OE , excluding end-points, is

Figure 4. The space $\Delta = \tau(\Sigma_2)$

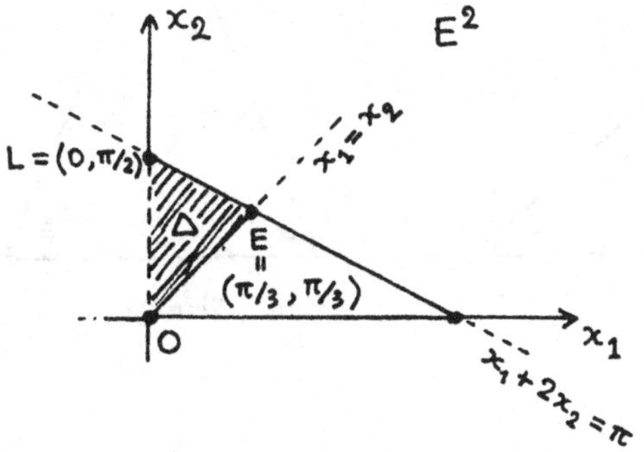

the image under τ of one component $\mathcal{I}_b$ of the symmetry type $\mathcal{I}$ of isosceles triangles $T(\alpha,\beta,\gamma)$, $0 < \alpha = \beta < \gamma$. The other component $\mathcal{I}_*$ of this symmetry type is mapped by τ to the open line-segment LE , and comprises the similarity classes of isosceles triangles $T(\alpha,\beta,\gamma)$ with $0 < \alpha < \beta = \gamma$. We refer to these two kinds of isosceles triangles as flat and sharp respectively (see Figure 5). Thus an isosceles triangle is flat or sharp according as its apex angle is greater than or less than $\pi/3$. The remaining symmetry type of triangles is the singleton $\mathcal{E}$ of all equilateral triangles.

To obtain a picture of the 'closed' space $\overline{\Sigma}_3$, we consider the closure of Δ in E^2 and collapse the interval OL to a point. The latter may then be identified with the singleton $* = \Sigma_2$. Thus $\overline{\Sigma}_3$ is homeomorphic to the closed disc shown in Figure 6, with symmetry types as indicated. In Figure 7, a succession of representative triangles for points on the boundary of $\overline{\Sigma}_3$ is shown to illustrate the continuous change of shape that occurs during a 'voyage round the rim' of $\overline{\Sigma}_3$.

Figure 5. Flat and sharp isosceles triangles

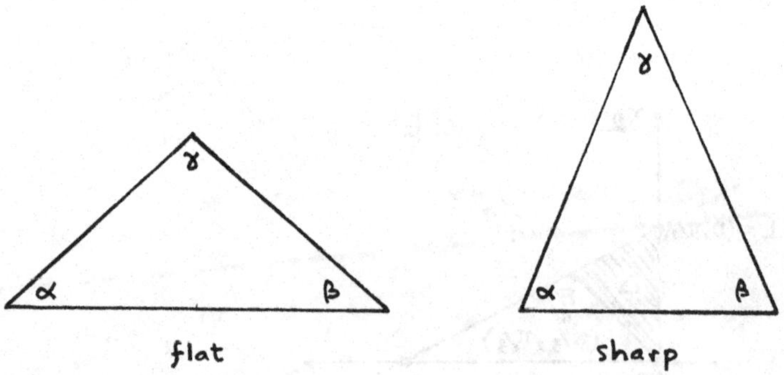

Exercise: Identify the set of similarity classes of right-angled triangles in $\bar{\Sigma}_3$ as an arc with end-points at $*$ and in $\mathcal{G}_\flat$.

Exercise: Show that the polarity map π_2 sends $\mathcal{E}$ and $\mathcal{L}$ to themselves and interchanges $\mathcal{G}_\flat$ and $\mathcal{G}_\#$.

Figure 6. <u>The space $\bar{\Sigma}_3$</u> .

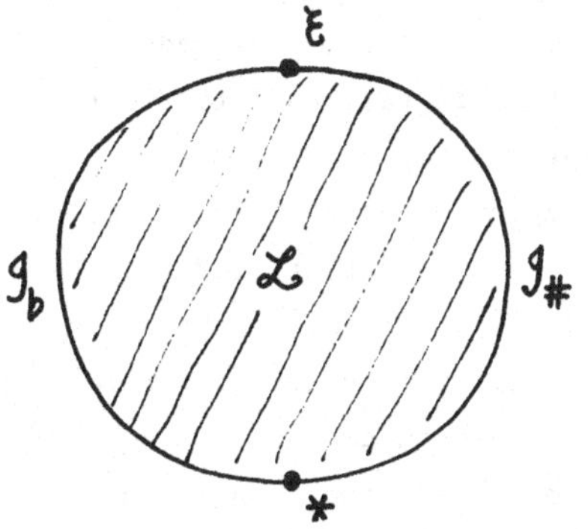

Figure 7. <u>A voyage round the rim of $\bar{\Sigma}_3$</u> .

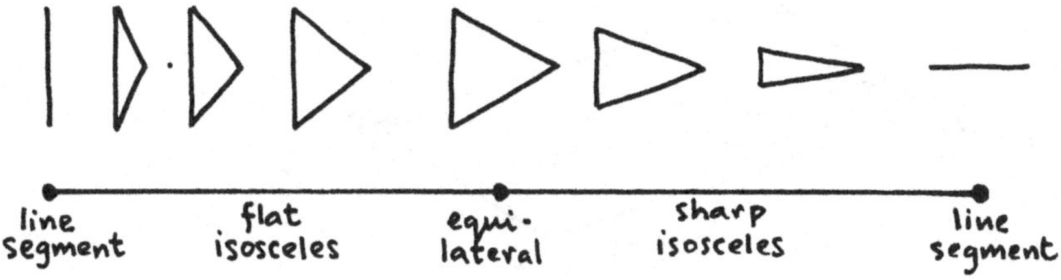

6. Quadrilaterals

Although the symmetry type $\mathcal{Q}$ of quadrilaterals having trivial symmetry is 4-dimensional, all other symmetry types of quadrilaterals are of dimension 2 at most. We should expect, therefore, to be able to give a full description of that part of $\overline{\Sigma}_4$ corresponding to 'symmetric' quadrilaterals. Let us denote the space $\overline{\Sigma}_4 \setminus \mathcal{Q}$ by $\mathcal{Q}^*$. Thus $\mathcal{Q}^*$ is made up of the space $\overline{\Sigma}_3$ described in §4 and the space of similarity classes of all quadrilaterals Q for which $\Gamma(Q)$ is nontrivial.

To complete the description of $\mathcal{Q}^*$, we examine the six symmetry types of symmetric quadrilaterals in the manner of §4. We discuss the symmetry type of parallelograms and its frontier in $\overline{\Sigma}_4$ in some detail to illustrate the procedure.

Let P be a parallelogram with vertices A, C, B, D in cyclic order. Suppose that $AB > CD$ and that $C\hat{A}B = \alpha > \beta = CB\hat{A}$. Then the similarity class of P is specified by the point $(\alpha, \beta) \in E^2$. Let us denote this class by $P(\alpha, \beta)$. Thus if $\mathcal{P}$ denotes the symmetry type of parallelograms, then there is an embedding $p: \mathcal{P} \to E^2$ given by $p(P(\alpha,\beta)) = (\alpha,\beta)$. The image of $\mathcal{P}$ is the open triangular region

$$\mathcal{P}^* = \{(x_1, x_2) \in E^2 : 0 < \beta < \alpha , \quad \alpha + \beta < {}^{\pi}/_{2}\} .$$

Figure 8. Parametrisation of parallelograms

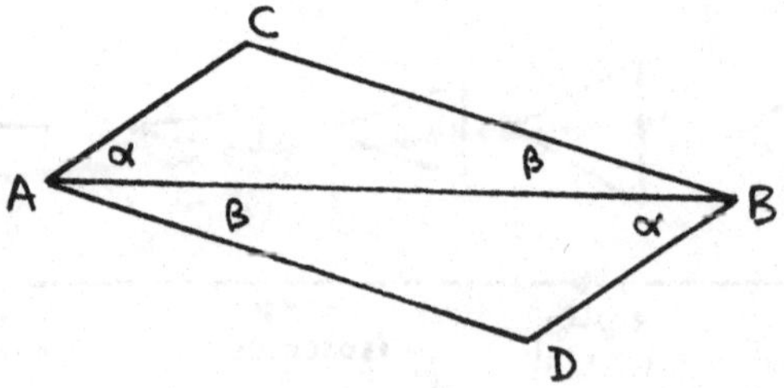

There are three symmetry types of quadrilaterals in the frontier of $\mathcal{P}$. These are the 1-dimensional type $\mathcal{O}$ of rectangles or oblongs given by $0 < \beta < \alpha$, $\alpha + \beta = \pi/2$, the 1-dimensional type $\mathcal{D}$ of rhombuses or diamonds given by $0 < \alpha = \beta$, $\alpha + \beta < \pi/2$, and the 0-dimensional type $\mathcal{S}$ of squares, given by $\alpha = \beta = \pi/4$. The closure of $\mathcal{P}$ in $\overline{\Sigma}_4$ is homeomorphic to the disc obtained from the closure of $\mathcal{P}^*$ in E^2 by collapsing the vertical side $\alpha = 0$ of $\mathcal{P}^*$ to a point, the type * of line-segments.

Figure 9. Parallelograms

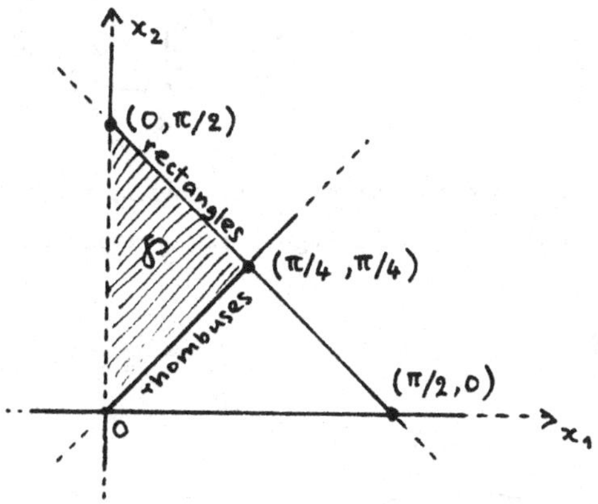

Figure 10. Parallelograms and their neighbours

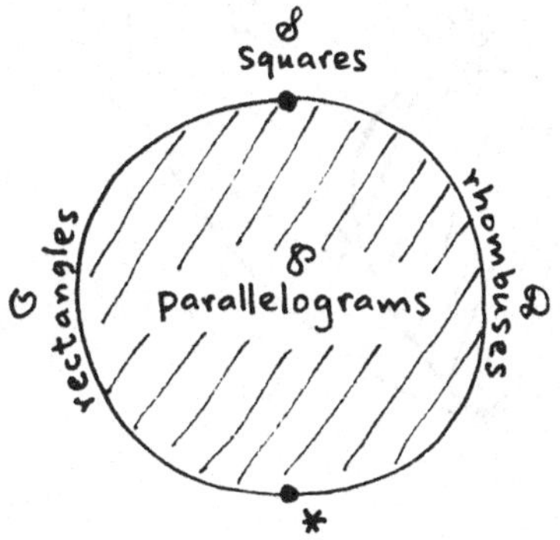

This procedure can be repeated for the other two 2-dimensional symmetry types (kites and isosceles trapezia). Their closures in $\overline{\Sigma}_4$ are homeomorphic to the discs shown in Figure 11.

Figure 11. Kites and Trapezia

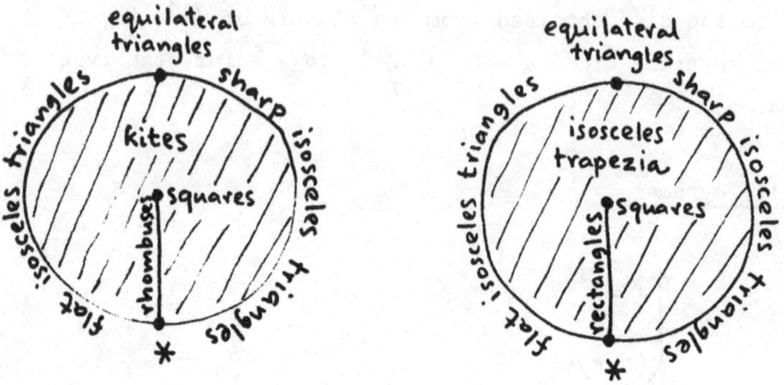

We may then fit these discs together along their common edges, to obtain the 2-dimensional cell-complex shown in Figure 12 as embedded in E^3 .

Figure 12. The space $\mathcal{Q}^*$

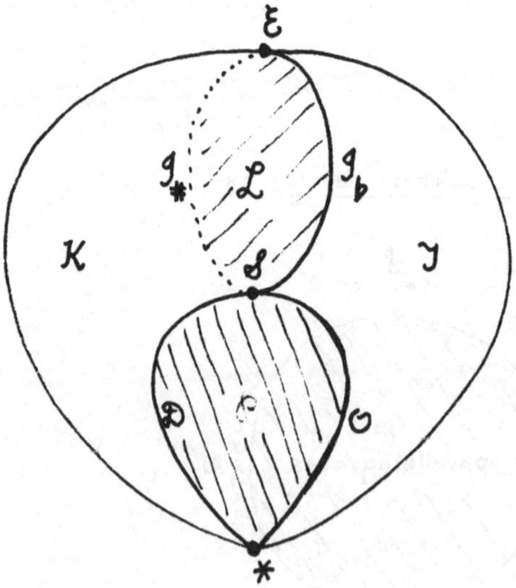

Exercise: The bilateral symmetry of Figure 12 is not accidental. Show that the polarity map π_2 fixes $\mathcal{E}$, $\mathcal{S}$, $*$, $\mathcal{P}$ and $\mathcal{L}$, interchanges $\mathcal{I}_*$ with $\mathcal{I}_b$, $\mathcal{D}$ with $\mathcal{O}$, and $\mathcal{K}$ with $\mathcal{T}$.

The above exercise is a simple extension of the Exercise at the end of §4. What we are beginning to see emerge is the bilateral symmetry of the space Σ as a whole, or, more generally, of $\mathfrak{S}$ itself.

7. The 1-skeleton

If P is any polygon, then $\operatorname{def} P = \alpha + \beta - \xi - 2$, according to the Proposition of §4, where α is the number of vertex-orbits, β is the number of edge-orbits, and ξ the dimension of the fixed point set in aff P, for the symmetry group $\Gamma(P)$. Since $\alpha \geqslant 1$ and $\beta \geqslant 1$, $\operatorname{def} P = 0$ iff $\alpha = \beta = 1$ and $\xi = 0$. It follows that P is perfect iff it is regular.

Likewise, $\operatorname{def} P = 1$ iff $\alpha + \beta = \xi + 3$ and the only solutions are

(i) $(\alpha,\beta,\xi) = (1,2,0)$

(ii) $(\alpha,\beta,\xi) = (2,1,0)$

(iii) $(\alpha,\beta,\xi) = (2,2,1)$.

Case (i) characterises the family of semi-regular or 0-regular polygons, and case (ii) the family of 1-regular polygons. Case (iii) characterises the symmetry type of isosceles triangles.

For each $n \geqslant 3$, there is a connected 1-dimensional symmetry type σ_n of 0-regular 2n-gons, and a connected 1-dimensional symmetry type σ_n^* of 1-regular 2n-gons, the end-points of both σ_n and σ_n^* being the nodes ρ_n and ρ_{2n} of regular n-gons and regular 2n-gons. For $n = 2$ and $n = 3$, rectangles, diamonds and the two arcs of isosceles triangles also supply 1-dimensional types terminating at the node $* = \rho_2$.

In this way we find that the 1-skeleton of the space Σ is composed of countably many 'daisy chains', each of which contain countably many symmetry types. Each chain is invariant under polarity, which fixes nodes and interchanges the two arcs of each loop.

Figure 13. The 1-skeleton of Σ .

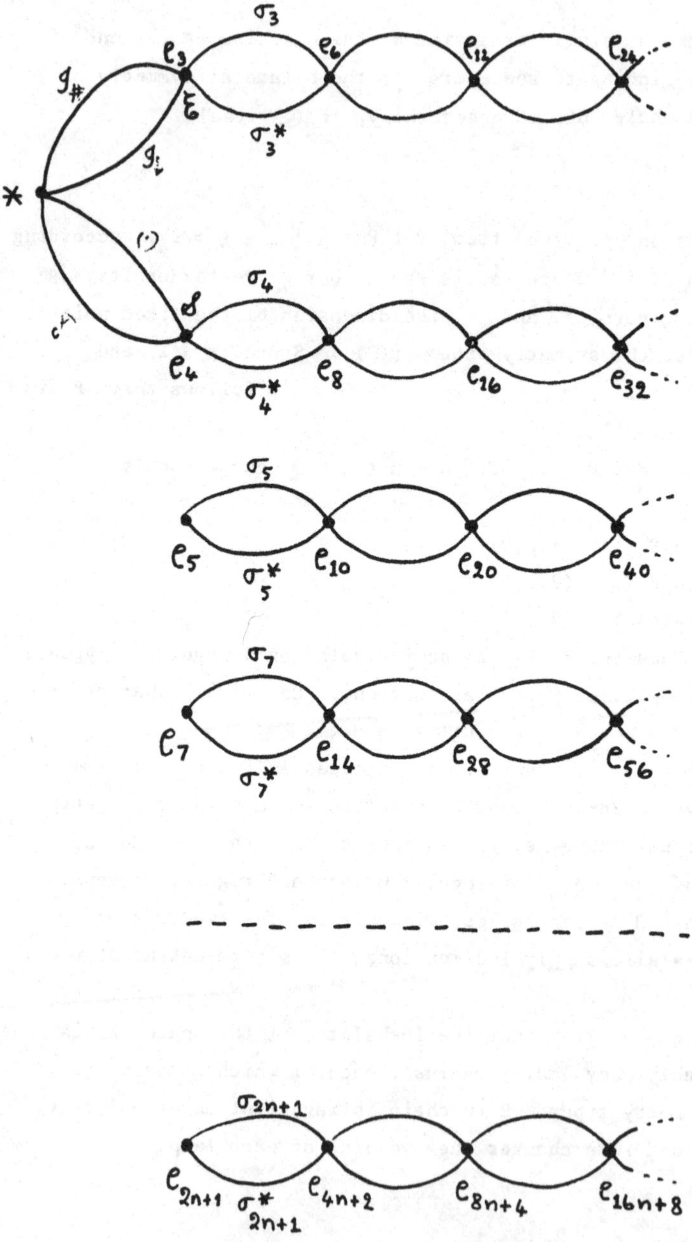

Some representative polygons in the first four loops are indicated in Figure 14. These chains will be used in the next chapter where we try to understand the symmetry type structure of vertex-regular prisms and their duals.

Figure 14: The first four loops

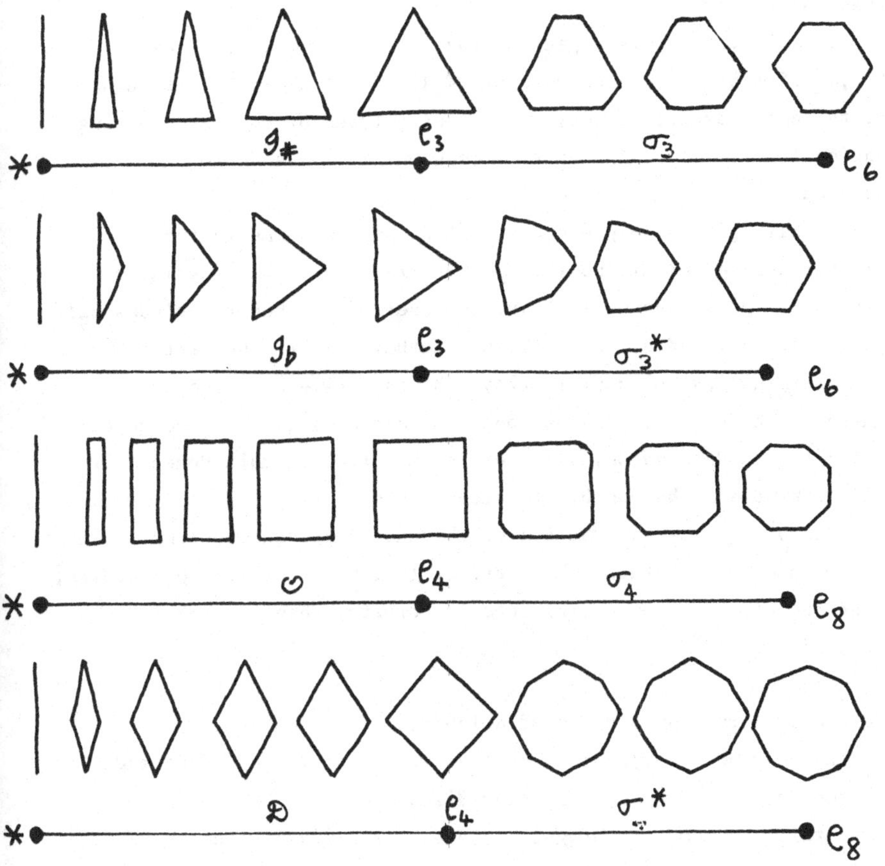

6. POLYHEDRA

For any given polyhedron P, the symmetry type structure of $\mathbb{C}_*P/\sim$ can be worked out in some detail, given enough time and patience. The dimension of each symmetry type can be determined by *ad hoc* methods and checked against Deicke's conjecture. More sophisticated topological invariants are more inaccessible: little is known about the relation between elementary topological invariants of $\mathbb{X}P$ (the Betti numbers for example) and the action of $\Gamma(P)$ on $F(P)$, although the latter determines the former.

To illustrate this aspect of the theory, we take the case $P = \square_3$, referring to the members of $\mathbb{C}_*\square_3$ as 'cuboids'. It turns out that there are exactly twenty two symmetry types of cuboids, ranging in dimension from 0 to 11, with dimensions 8, 9 and 10 unrepresented.

We also give a complete description of the symmetry type structure for the set of all polyhedra that are either face-regular or vertex-regular. As we showed in §2.9, this set of polyhedra includes all the edge-regular polyhedra, which are none other than the perfect polyhedra, forming just nine similarity classes. Five of these nodes are represented by the Platonic solids, and the remaining four by the cuboctahedron, the icosidodecahedron, and their polars (the rhombic dodecahedron of the first kind and the rhombic triacontahedron).

One of the most attractive of the many unsolved problems is the status of Deicke's conjecture, which states that if P is any polyhedron, and the action of $\Gamma(P)$ on $F_1(P)$ has ε orbits, then P has deficiency $\varepsilon - 1$.

1. Some combinatorial properties of polyhedra

Let P be a polyhedron heaving v vertices, e edges and f faces. Then $v - e + f = 2$, as Euler discovered in 1750. For the history of this theorem, see Biggs, Lloyd & Wilson [1976], Lakatos [1976] and Pont [1974]. A much more obvious fact is that the multiplicity of P is given by $\mu(P) = 2e$. To see this, let η be any edge of P. Then η is incident with two vertices of P, say ξ and ξ', and with two faces of P, say ζ and ζ'. Hence for each edge η there are four incident pairs $(\xi,\xi),(\xi',\xi),(\xi,\zeta')$ and (ξ',ζ'). Each of these pairs occurs in the corresponding set of four pairs for just one other edge of P.

For example, in Figure 1, (ξ,ζ) is one of the four pairs associated with the edge η'. It follows that $\mu(P) = \frac{1}{2}(4e) = 2e$. For instance, the cube $\square_3$ has 12 edges and multiplicity 24.

From the Euler relation and the expression just obtained for the multiplicity, we can deduce a simplified version of the formula for dimension of combinatorial types given in §2.3. For any polyhedron P,

$$\begin{aligned} \dim C_*P &= 3(f_o(P) + f_2(P)) - \mu(P) \\ &= 3(v + f) - 2e \\ &= 3(2 + e) - 2e \\ &= e + 6 . \end{aligned}$$

One immediate consequence is that, for any polyhedron P, $\operatorname{def} P \leq e - 1$, since the dimension of the Euclidean group $\mathrm{Iso}(e)$ is 7. For instance, $\dim C_* \square_3 = 18$ and for all $P \overset{\sim}{\sim} \square_3$, $\operatorname{def} P \leq 11$.

2. Cuboids

A polyhedron P such that $P \overset{\sim}{\sim} \square_3$ will be called a <u>cuboid</u>. Thus any cuboid has six faces, each of which is a quadrilateral, and each of its eight vertices is incident with just three of these faces. In order to determine the various symmetry types of cuboids, we must study the action of $\Gamma(P)$ on the face-lattice $F(P)$, for each cuboid P.

In fact, the induced action of $\Gamma(P)$ on the set $F_2(P)$ of faces of P gives a faithful representation of $\Gamma(P)$ as a subgroup of

Figure 1 <u>Incidence in polyhedra</u>

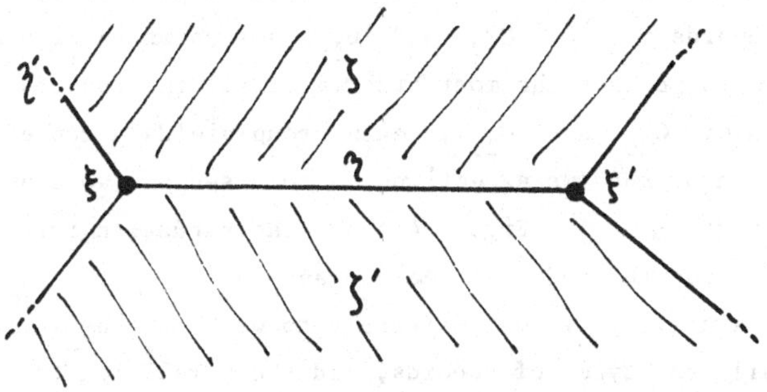

the permutation group S_6 . Let us label the faces of a cuboid P as A, $\bar{A}$, B, $\bar{B}$ and C, $\bar{C}$, where X and $\bar{X}$ have no vertices in common. Thus X and $\bar{X}$ are 'opposite' faces, X = A, B or C . This notation allows us to specify the various possible symmetry group actions in very concise form, up to conjugacy in the combinatorial automorphism group Aut P . The latter is a group of order 48 . It is important to realise that conjugacy within Aut P is more restrictive than conjugacy within Iso(3) . That is, two cuboids (and, more generally, two combinatorially equivalent polyhedra) need not be symmetry equivalent even though their symmetry groups are conjugate to one another in Iso(3)).

In Table 2 are listed the 22 conjugacy classes of subgroups of Aut $\square_3$ that occur as the symmetry group of some cuboid. That is to say, if P is any cuboid then there is a lattice isomorphism $\lambda:F(P) \to F(\square_3)$ that induces a group isomorphism from $\Gamma(P)$ onto G_i , $1 \leq i \leq 22$. Note the inverse correlation between the order of G_i and the dimension of the associated symmetry type. In all, Aut $\square_3$ has 33 conjugacy classes of subgroups. Thus 11 conjugacy classes are not represented by the symmetry group of a cuboid. For example, there are three classes of subgroups of order 24 , while no cuboid has symmetry group of this order.

A rewarding exercise for those who like to study physical representations of geometrical objects is to make solid models of all 22 types of cuboids, either by cutting them from solid blocks of wood, or by glueing laminae edge to edge. Many types of cuboids occur in everyday life in the shape of familiar objects, while others may seem very strange indeed. In particular, I have never come across natural or man-made objects in the form of cuboids with group G_2, G_3, G_7, G_8, G_{14} or G_{15} . To help the reader in the task of visual realisation, pictures of the cuboids for G_3, G_{17} and G_{19} are shown in Figure 3. The first of these is perhaps the most bizarre of all the cuboids. With the exception of G_{20} and G_{21} , each group yields a connected symmetry type. Each of the two exceptions is composed of two open arcs separated by the node ※$\square_3$. Figure 4 shows the situation for G_{21} . The type of G_{20} joins the nodes * , ※$\square_3$ and π_6 .

More generally, it is not difficult to work out the incidence relations among all 22 types of cuboids, and their relation to the types that occur as limits of sequences of cuboids (tetrahedra, triangles, quadrilaterals, etc.).

Table 2 <u>The symmetry types of cuboids</u>

Label	Presentation	Order	Deficiency
G_1	-	1	11
G_2	$\langle (A\bar{A})(B\bar{B}) \rangle$	2	5
G_3	$\langle (AB)(\bar{A}\bar{B})(C\bar{C}) \rangle$	2	6
G_4	$\langle (A\bar{A})(B\bar{B})(C\bar{C}) \rangle$	2	5
G_5	$\langle C\bar{C} \rangle$	2	7
G_6	$\langle (A\bar{B})(\bar{A}B) \rangle$	2	6
G_7	$\langle (ABC)(\bar{A}\bar{B}\bar{C}) \rangle$	3	3
G_8	$\langle G_2, G_3 \rangle$	4	3
G_9	$\langle G_2, G_5 \rangle$	4	3
G_{10}	$\langle (A\bar{A}), (B\bar{B}) \rangle$	4	4
G_{11}	$\langle G_2, G_6 \rangle$	4	3
G_{12}	$\langle G_3, G_4 \rangle$	4	3
G_{13}	$\langle G_3, G_5 \rangle$	4	4
G_{14}	$\langle (ABC)(\bar{A}\bar{B}\bar{C}), (A\bar{A})(B\bar{B})(C\bar{C}) \rangle$	6	2
G_{15}	$\langle (ABC)(\bar{A}\bar{B}\bar{C}), (BC)(\bar{B}\bar{C}) \rangle$	6	2
G_{16}	$\langle (A\bar{A})(B\bar{B}), (A\bar{A})(C\bar{C}), G_4 \rangle$	8	2
G_{17}	$\langle G_8, G_4 \rangle$	8	2
G_{18}	$\langle G_8, (B\bar{B}) \rangle$	8	2
G_{19}	$\langle (AB\bar{A}\bar{B}), (B\bar{B}) \rangle$	8	2
G_{20}	$\langle (ABC)(\bar{A}\bar{B}\bar{C}), (A\bar{A})(B\bar{C})(\bar{B}C), G_4 \rangle$	12	1
G_{21}	$\langle (AB\bar{A}\bar{B}), (AB)(\bar{A}\bar{B})(C\bar{C}), G_4 \rangle$	16	1
G_{22}	Aut $\square_3$	48	0

Figure 3 Three kinds of cuboids

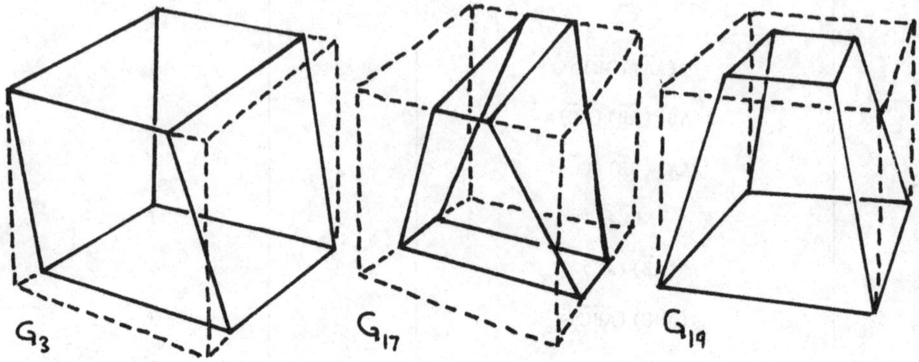

Figure 4 Prisms on a square base

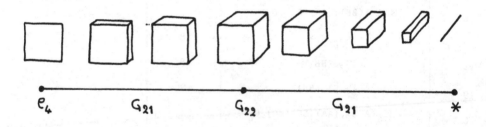

By polarity, we obtain automatically the corresponding facts about polyhedra $Q \approx \diamondsuit_3$. Again, it may be a good idea to construct physical models to appreciate the full variety of shapes that are possible.

3. Vertex-regular polyhedra

The above discussion of cuboids lacks full detail because of the high dimension and topological complexity of many of the symmetry types that occur. When we switch attention to the family of vertex-regular polyhedra, however, we encounter a far simpler set of problems, since none of the symmetry types that occur has dimension greater than 2, and all but a few are connected and simply-connected. In fact, we can describe the space of similarity classes of such polyhedra, and its symmetry type structure, in full detail by piecing together a sequence of 2-dimensional cell-complexes, each of which can be easily embedded in E^3 .

The bulk of this space (all but one of the complexes in our sequence) is composed of the countable infinity of types of vertex-regular prisms. In this context, we say that a vertex-regular polyhedron in E^3 is a <u>prism</u> iff its symmetry group keeps some line in E^3 setwise fixed.

The non-prisms or 'isotropic' vertex-regular polyhedra form the remaining part of the space, and is made up of 23 symmetry types. It is interesting to find that five of these types are the nodes represented by the five Platonic solids, and thirteen others contain the similarity classes of the Archimedean solids (other than prisms). However, only two of these thirteen types are nodes, the others having dimension 1 or 2 . Moreover, there are five further symmetry types that do not include a similarity class of Platonic or Archimedean solids. To my mind, these are equally worthy of attention; I am particularly fond of the polyhedra whose types are labelled J and K (components of one type) and U below. See Figure 10 and Tables 12(a) and 13.

We now explain how the space of vertex-regular polyhedra can be explored. Let P be a vertex-regular polyhedron. Then P is similar to a normal polyhedron in E^3 and we may suppose at the outset that P itself is such a polyhedron. Thus every vertex of P lies on the unit sphere S^2 in E^3, and the symmetry group $G = \Gamma(P)$ of P is a finite subgroup of $O(3)$. Moreover, vert P is the G-orbit $G\cdot v$ of any vertex v of P.

Conversely, for any point p of S^2 and any finite subgroup H of $O(3)$, the H-orbit $H\cdot p$ of p is the vertex set of some n-polytope P, where $0 \leq \dim P \leq 3$. To guarantee that $\dim P = 3$, we need to ensure that p does not lie on any proper linear subspace of E^3 that is setwise fixed under the action of H.

If $P = \operatorname{conv} H\cdot p$ is a polyhedron, then H is a subgroup of $\Gamma(P)$, and may well be a proper subgroup of $\Gamma(P)$. For instance, P could be a cube and H the group of rotational symmetries of P.

These remarks show that for each nontrivial finite subgroup G of $O(3)$, there is a continuous map

$$\theta_G : S^2 \to \mathfrak{S}(3)$$

sending $x \in S^2$ to the similarity class of the convex hull of $G\cdot x$. Moreover, the union of the images $\theta_G(S^2) = \tilde{G}$ is that portion $\mathfrak{J}(3)$ of $\mathfrak{S}(3)$ made up of similarity classes of vertex-transitive polyhedra, vertex-transitive polygons and 1-polytopes.

To describe $\mathfrak{J}(3)$, it is enough to describe $\tilde{G}$ for some group G in each conjugacy class of nontrivial finite subgroups of $O(3)$. The space $\mathfrak{J}(3)$ is then assembled by identifying the various images $\tilde{G}$ along common symmetry types. Since S^2 is a surface, the symmetry types in each $\tilde{G}$ and hence in $\mathfrak{J}(3)$ are all of dimension ≤ 2. That is to say, for any vertex-regular polyhedron P, $\operatorname{def} P \leq 2$.

As an illustration of the procedure, consider the group $G = \Gamma(\square_3)$ of all symmetries of the cube $\square_3$. This is a group of order 48, consisting of all matrices of the form AJ, where J is any one of the eight diagonal matrices $\operatorname{diag}(\pm1, \pm1, \pm1)$ and A is any of the six permutation matrices of order 3×3. For example, we could take

$$A = \begin{bmatrix} 0 & 1 & 0 \\ 1 & 0 & 0 \\ 0 & 0 & 1 \end{bmatrix}, \quad J = \begin{bmatrix} -1 & 0 & 0 \\ 0 & 1 & 0 \\ 0 & 0 & -1 \end{bmatrix}$$

to obtain the element

$$AJ = \begin{bmatrix} 0 & 1 & 0 \\ -1 & 0 & 0 \\ 0 & 0 & -1 \end{bmatrix}$$

of G. In fact, G is generated by the three matrices

$$M_1 = \begin{bmatrix} -1 & 0 & 0 \\ 0 & 1 & 0 \\ 0 & 0 & 1 \end{bmatrix}, \quad M_2 = \begin{bmatrix} 0 & -1 & 0 \\ -1 & 0 & 0 \\ 0 & 0 & 1 \end{bmatrix} \quad \text{and} \quad M_3 = \begin{bmatrix} 1 & 0 & 0 \\ 0 & 0 & 1 \\ 0 & 1 & 0 \end{bmatrix}$$

which are reflexions in the planes $x_1 = 0$, $x_1 + x_2 = 0$ and $x_2 = x_3$ respectively. Figure 5 shows a division of the surface K of the cube $\square_3$ into 48 triangular regions, each of which is the intersection with this surface of a fundamental region of G in E^3. The particular region D bounded by the planes of reflexion of M_1, M_2 and M_3 is shaded.

Figure 5 A fundamental region for $\Gamma(\square_3)$

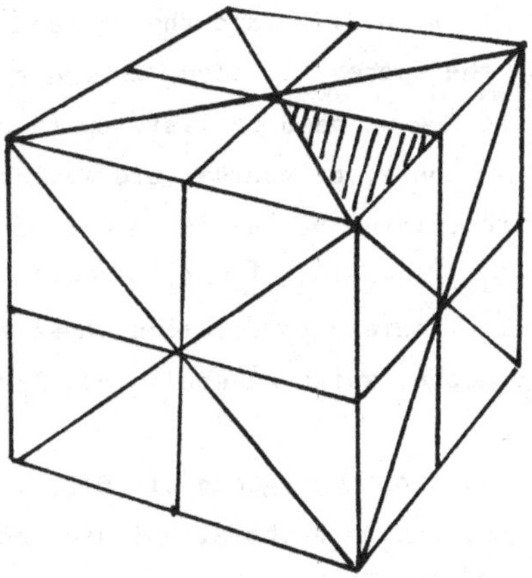

The image of K (or rather of its radial projection K' in S^2) under θ_G is the same as that of (the radial projection D' of) D. But D' is mapped homeomorphically into $\mathfrak{S}$ by θ_G, and its interior, its three edge arcs and its three corners are mapped onto symmetry types in $\mathfrak{S}$. Thus $\tilde{D}' = \tilde{K}'$ is composed of one 2-dimensional symmetry type, three types of dimension 1 and three nodes. Representative polyhedra of these seven types are easily constructed by using the fact that for any $x = (x_1,x_2,x_3) \in D'$, the point $\theta_G(x)$ in $\mathfrak{S}$ is represented by the polyhedron whose vertices are the 48 points (some of which may coincide with one another)

$$(\pm x_1, \pm x_2, \pm x_3), \quad (\pm x_1, \pm x_3, \pm x_2), \quad (\pm x_3, \pm x_1, \pm x_2)$$
$$(\pm x_2, \pm x_1, \pm x_3), \quad (\pm x_2, \pm x_3, \pm x_1), \quad (\pm x_3, \pm x_2, \pm x_1).$$

Seven such polyhedra are shown in Figure 6. For convenience, each is shown as conv $G \cdot x$, where $x \in D$. The coordinates of x are indicated in each case, together with the deficiency d. The capital letters attached to each type are those used as labels by Robertson and Carter [1970]. We may sum up these findings by saying that the group $\Gamma(\square_3)$ contributes a 2-dimensional region to $\mathfrak{S}$, homeomorphic to a closed disc, as shown in Figure 7. This disc is partitioned into seven symmetry types of dimensions 0, 1 and 2; the labels may be used to match the types in Figure 7 to the polyhedra of Figure 6.

The procedure that we have just described may be repeated for every nontrivial finite subgroup of O(3), and the space $\mathfrak{J}(3)$ can then be put together by identifying the spaces $\tilde{G}$ along common symmetry types. To carry out this scheme in practice we need to list, up to conjugacy, all the finite subgroups of O(3). These, of course, are well-known, and many detailed accounts of how to derive such a list are readily available. See, for example, Coxeter [1948], Fejes Tóth [1964] or Weyl [1952]. For present purposes, it is sufficient merely to describe these groups. They fall naturally into two main classes, which we shall call 'prismatic' and 'nonprismatic'.

Let G be a nontrivial finite subgroup of O(3). Then G fixes the origin O and E^3 setwise as a whole, but need not fix setwise any proper linear subspace of E^3. If there is such a proper linear subspace A for G, then G is said to be <u>prismatic</u> and A is called an <u>axis</u> of G. Otherwise G is nonprismatic.

Figure 6   Vertex-regular polyhedra with group $\Gamma(\square_3)$

$x = (1,1,1), d = 0.$ $x = (0,0,1), d = 0.$ $x = (0,1,1), d = 0.$

$x = (\lambda,\lambda,1), d = 1,$ $x = (\lambda,1,1), d = 1$ $x = (0,\lambda,1), d = 1,$
$0 < \lambda < 1.$ $0 < \lambda < 1.$ $0 < \lambda < 1.$

$x = (\lambda,\mu,1), d = 2,$
$0 < \lambda < \mu < 1.$

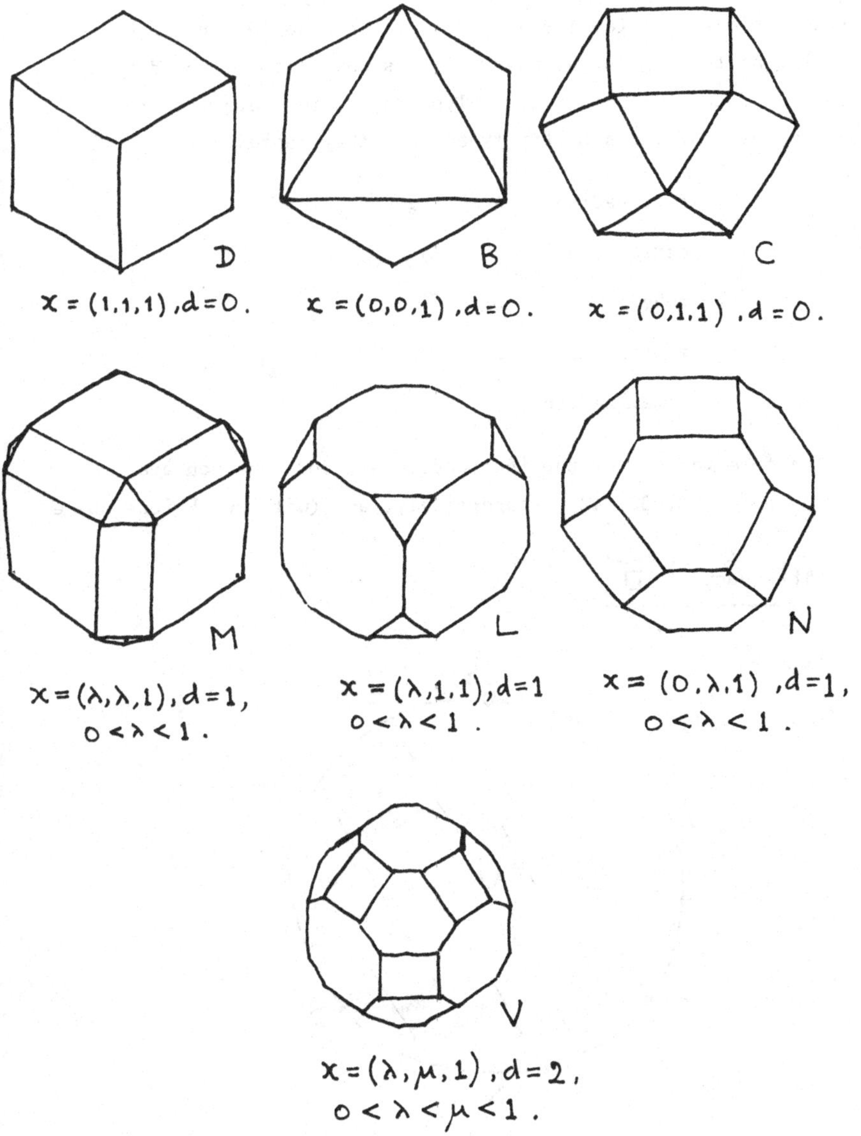

4. The nonprismatic groups

We approach these groups via the Platonic solids. The latter form a family of five (similarity classes of) polyhedra, and are so named because they are discussed by Plato in his dialogue Timaeus (among English translations, see Lee [1965]). The Platonic solids are of course the same thing as the regular polyhedra. Thus any Platonic solid is similar to one of the following five polyhedra, listed here with the labels that we use for the symmetry types that they represent.

tetrahedron	A	Δ_3
octahedron	B	$\Diamond_3$
cube	D	$\Box_3$
icosahedron	E	
dodecahedron	G	

The face vectors of the icosahedron and dodecahedron are (12, 30, 20) and (20, 30, 12) respectively, and (with an obvious abuse

Figure 7 The space $\widetilde{\Gamma(\Box_3)}$

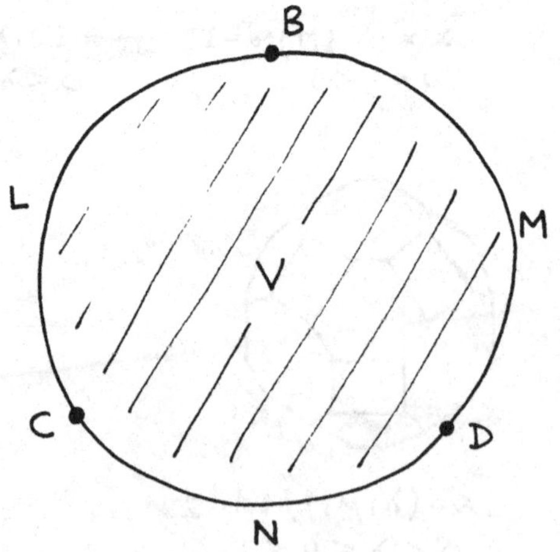

of notation) $E^* \sim G$. Figure 8 indicates how an icosahedron may be embedded in a cube, and a cube in a dodecahedron. Given that the cube has vertices $(\pm 1, \pm 1, \pm 1)$, we may take one vertex of the icosahedron to be $(0,\lambda,1)$ and of the dodecahedron $(\lambda,0,\mu)$, where $\lambda = \frac{1}{2}(\sqrt{5}-1)$ and $\mu = 1 - \frac{1}{2}\,(6-2\sqrt{5})$.

Suppose then that P is a Platonic solid in E^3 with centroid O , and let $G = \Gamma(P)$. Then G has a subgroup $G_+ = \Gamma_+(P)$ consisting of all the rotational symmetries of P . Thus $g \in G_+$ iff $g \in G$ and $\det g = 1$. The subgroup G_+ has index 2 in G . Let $j \in O(3)$ be the matrix

$$j = \begin{bmatrix} -1 & 0 & 0 \\ 0 & -1 & 0 \\ 0 & 0 & -1 \end{bmatrix}$$

which acts on E^3 by reflexion in O . Then every nonprismatic subgroup of $O(3)$ is conjugate in $O(3)$ to one of the groups

$$G \ , \quad G_+ \ , \quad \langle j, G_+ \rangle \ .$$

Figure 8 Relations between the cube, icosahedron and dodecahedron

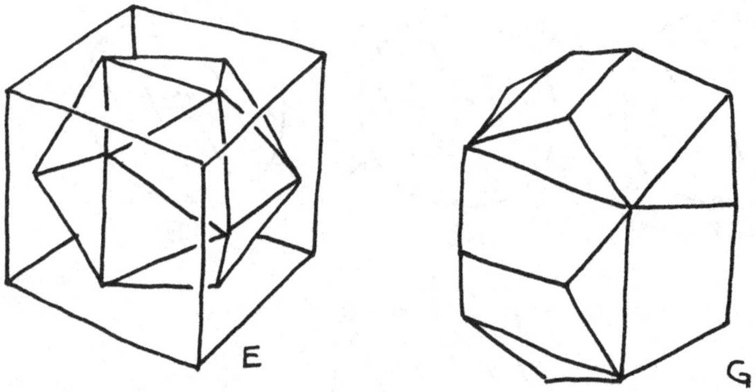

At first sight, this yields no less than 15 conjugacy classes. However, since $\langle j,G_+\rangle = G$ except for the case $P = \Delta_3$, and since $\Gamma(P) = \Gamma(P^*)$, there are only seven conjugacy classes of nonprismatic groups, namely those of

$$\Gamma(A)\ \Gamma(B),\ \Gamma(E),\ \Gamma_+(A),\ \Gamma_+(B),\ \Gamma_+(E)\ \text{ and }\ \langle j,\Gamma_+(A)\rangle\ .$$

These groups have orders 24, 48, 120, 12, 24, 60 and 24 respectively, and each group yields a 2-dimensional disc of similarity classes of polyhedra in the space $\mathfrak{S}$, as described already for the case $\Gamma(B)$. The full set of seven discs is shown in Figure 9.

Figure 9 Symmetry types of the nonprismatic groups

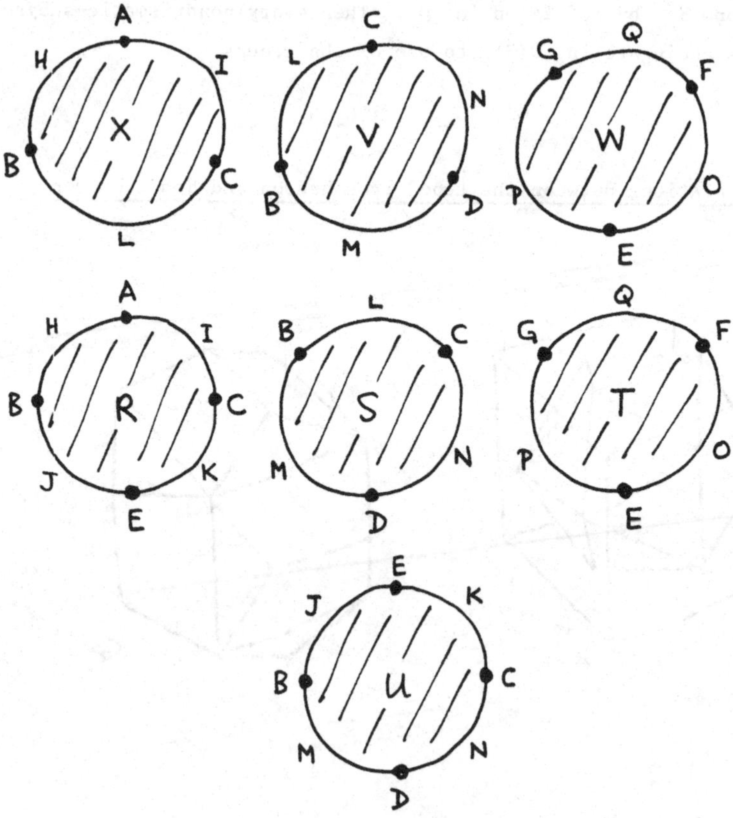

A glance at Figure 9 reveals that every bounding arc occurs in more than one disc. Thus if we make the appropriate identifications along these arce, we obtain a connected 2-dimensional cell-complex. It turns out that this complex can be embedded in E^3 and such an embedding is sketched in Figure 10.

The next three Tables 11, 12 and 13 are designed to display representative polyhedra of the 24 cells in the complex of Figure 10, and to relate this family to the Platonic and Archimedean solids. All but six of the cells are represented by a Platonic or an Archimedean solid, and these are marked by asterisks and daggers respectively. Of the six 'strangers', J and K are components of the same symmetry type. In Tables 12 and 13, none of the polyhedra shown are Archimedean, some of their faces being semi-regular rather than regular in each case. Thus Archimedean solids of deficiency 1 or 2 are not 'generic' representatives of their symmetry types: the fact that all the faces can be chosen to be regular is a 'fluke'.

To assist in relating the information in the tables to the topology of the complex, Figure 14 shows the variation in the polyhedral structure from D along N to C .

Figure 10 The complex of nonprismatic vertex-regular polyhedra

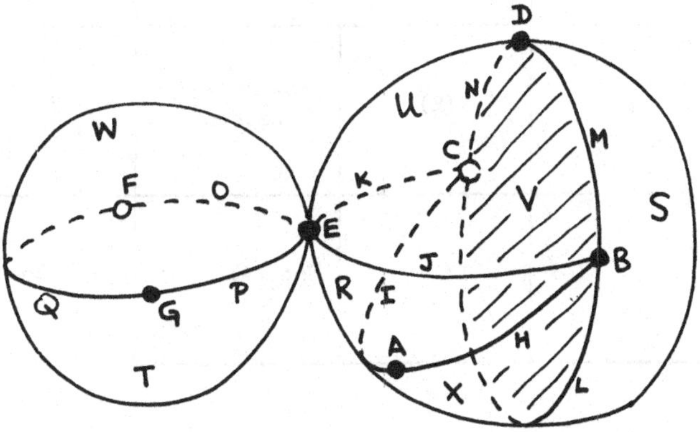

Table 11 Vertex-regular polyhedra of deficiency 0

	LABEL	SYMMETRY GROUP	CLASSICAL NAME	CLASSICAL SYMBOL
	A *	Γ(A)	tetrahedron	333
	B *	Γ(B)	octahedron	3333
	C †	Γ(B)	cuboctahedron	3434
	D *	Γ(B)	cube	444
	E *	Γ(E)	icosahedron	33333
	F †	Γ(E)	icosidodecahedron	3535
	G *	Γ(E)	dodecahedron	555

Table 12(a) Vertex-regular polyhedra of deficiency 1

	LABEL	SYMMETRY GROUP	CLASSICAL NAME	CLASSICAL SYMBOL
	H †	$\Gamma(A)$	truncated tetrahedron	366
	I	$\Gamma(A)$	-	-
	J	$\Gamma_*(A)$	-	-
	K	$\Gamma_*(A)$	-	-
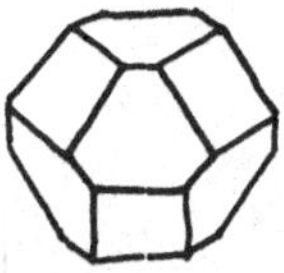	L †	$\Gamma(B)$	truncated octahedron	466

Table 12(b) Vertex-regular polyhedra of deficiency 1

	LABEL	SYMMETRY GROUP	CLASSICAL NAME	CLASSICAL SYMBOL
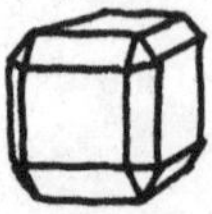	M †	Γ(B)	rhombi-cuboctahedron	3444
	N †	Γ(B)	truncated cube	388
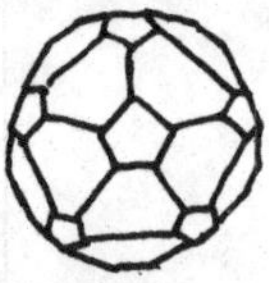	O †	Γ(E)	truncated icosahedron	566
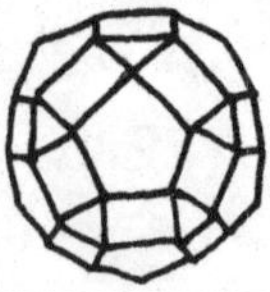	P †	Γ(E)	rhomb-icosidodecahedron	3454
	Q †	Γ(E)	truncated dodecahedron	31010

Table 13 Vertex-regular polyhedra of deficiency 2

	LABEL	SYMMETRY GROUP	CLASSICAL NAME	CLASSICAL SYMBOL
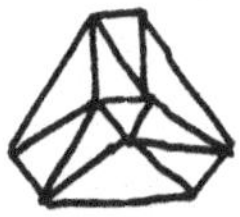	R	$\Gamma_{+}(A)$	-	-
	S †	$\Gamma_{+}(B)$	snub cube	33334
	T †	$\Gamma_{+}(E)$	snub dodecahedron	33335
	U	$\Gamma_{*}(A)$	-	-
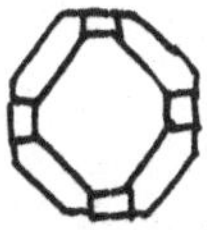	V †	$\Gamma(B)$	rhombitruncated cuboctahedron	468
	W †	$\Gamma(E)$	rhombitruncated icosidodecahedron	4610
	X	$\Gamma(A)$	-	-

Figure 14 An arc of vertex-regular polyhedra

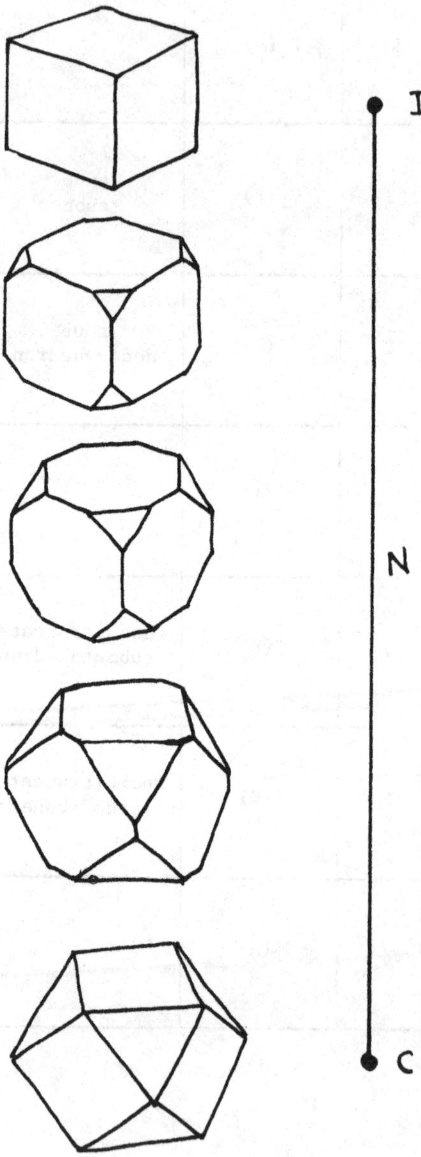

5. The prismatic groups

In addition to the seven conjugacy classes that we have referred to in §4, there are seven infinite sequences of conjugacy classes of non-trivial finite subgroups of $O(3)$. We shall adopt notations that conform fairly closely with established conventions. In particular, suppose that H is a finite subgroup of $SO(3)$ and that K is a subgroup of H of index 2. Thus if K has order n, then H has order $2n$. Since H is a rotation group, it does not contain the element j (reflexion in O). Now let K' denote the second coset $H \setminus K$ of K in H, and let jK' denote the set of all elements jk', where $k' \in K'$. Then the set $jK' \; K$ is a subgroup of $O(3)$, of order $2n$, and this group is denoted by HK. We have already encountered an example of this construction with $H = \Gamma(A)$ and $K = \Gamma_+(A)$. Then $\Gamma(A)\Gamma_+(A)$ is the group that we have denoted $\Gamma_*(A)$.

The seven sequences of classes may now be written

$$\mathcal{C}_m, \mathcal{D}_m, \bar{\mathcal{C}}_m, \bar{\mathcal{D}}_m, \mathcal{C}_{2m}\mathcal{C}_m, \mathcal{D}_{2m}\mathcal{D}_m \text{ and } \mathcal{D}_m\mathcal{C}_m ,$$

where in any integer $\geqslant 2$. These groups have orders $m, 2m, 2m, 4m, 2m, 4m, 2m$ respectively. The class $\mathcal{C}_m$ is that of the cyclic group of order m, generated by rotation through $2\pi/m$ about a line through O. The class $\mathcal{D}_m$ is that of the dihedral group generated by rotation through $2\pi/m$ about a line λ through O and rotation through π about a line μ through O perpendicular to λ. Likewise, a group of $\bar{\mathcal{D}}_m$ is generated by these two operations together with the reflexion j in O. For full details, see, for example, Fejes Tóth [1964], p. 65.

Only $\mathcal{D}_m$, $\bar{\mathcal{D}}_m$, and $\mathcal{D}_{2m}\mathcal{D}_m$ occur as the symmetry groups of vertex-regular polyhedra. There are, in fact, essentially five sequences of symmetry types of prismatic vertex-regular polyhedra, which we label C_m, D_m, E_m, F_m and G_m. These symbols are not related to the above group symbols, nor to those of §4. They are used here to maintain the conventions adopted by Robertson & Carter [1970], as follows:

label	description
C_m	right prisms on regular m-gons;
D_m	antiprisms on regular m-gons;
E_m	skew prisms on regular m-gons;
F_m	right prisms on semi-regular 2m-gons;
G_m	antiprisms on semi-regular 2m-gons;

Representative polyhedra of the five types are shown in Figure 15 for the $m = 5$ and $m = 3$. The deficiencies and symmetry groups are listed in Table 16. Notice that in four of the sequences there is a difference in the symmetry group class between m odd and m even.

The symmetry types of E_m, F_m and G_m are 2-cells and of C_m and D_m are open arcs, with the exception of C_4, D_2 and D_4 , each of which is composed of two open arcs, separated by the node D (cubes), A (tetrahedra) and B (octahedra) respectively. Figure 17 illustrates this phenomenon in the case of C_4 . Hence ρ_m is used as a label for the node of regular m-gons, and $* = \rho_2$ for the node of 1-polytopes. We also use σ_m to denote the 1-dimensional type of semi-regular 2m-gons. These labels are chosen to agree with the conventions of Chapter 5 .

We now find that the 'prismatic' part of $\mathcal{J}(3)$ is built by identifying the terms $\mathcal{K}_m$ of a sequence of 2-dimensional cell-complexes along their common cells, where $\mathcal{K}_m$ is the cell-complex shown in Figure 18. This picture is valid for all $m > 4$. For $m = 2, 3$ and 4 the picture requires slight modification to absorb the peculiarities discussed above. Note that the prismatic and nonprismatic parts of $\mathcal{J}(3)$ intersect in the three nodes A, B and D . Note also that part of the 1-skeleton of $\mathcal{J}(3)$ is formed by the chains of regular and semi-regular polygons discussed in Chapter 5. These types are indicated by the extra broken line in Figure 18.

Figure 15 <u>Some vertex-regular prisms</u>

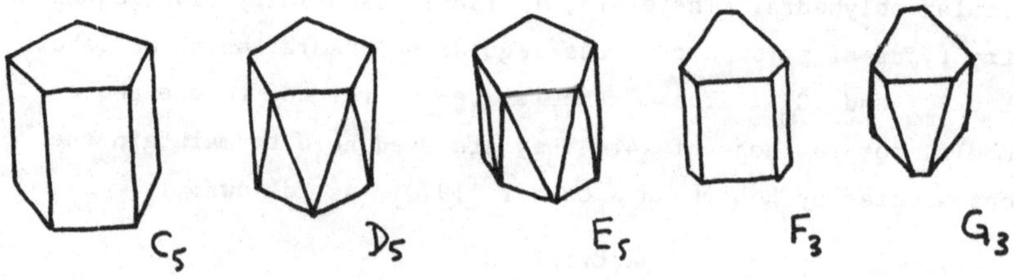

Table 16 Symmetry properties of vertex-regular prisms

LABEL	DEFICIENCY	SYMMETRY GROUP m even	m odd
C_m	1	$\bar{\mathcal{D}}_m$	$\mathcal{D}_{2m}\mathcal{D}_m$
D_m	1	$\mathcal{D}_{2m}\mathcal{D}_m$	$\bar{\mathcal{D}}_m$
E_m	2	$\mathcal{D}_m$	$\mathcal{D}_m$
F_m	2	$\bar{\mathcal{D}}_m$	$\mathcal{D}_{2m}\mathcal{D}_m$
G_m	2	$\mathcal{D}_{2m}\cdot\mathcal{D}_m$	$\bar{\mathcal{D}}_m$

Figure 17 The symmetry type C_4

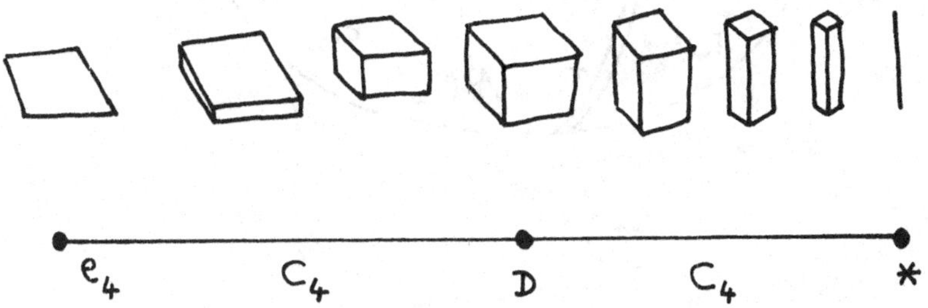

6. Face-regular polyhedra

Every statement about vertex-regular polyhedra translates automatically into a statement about face-regular polyhedra, by polarity. Thus we can deduce immediately from the results of the three preceding sections the detailed symmetry type structure of the space $\mathcal{J}^*(3)$ of face-regular polyhedra. It remains only to note that $\mathcal{J}(3)$ intersects $\mathcal{J}^*(3)$ in the sequence of nodes $\rho_m = A_m$ of regular polygons, and in the five nodes A, B, D E and G of the Platonic solids. The two nodes in $\mathcal{J}^*(3)$ that do not appear in $\mathcal{J}(3)$ are C^* and F^*, represented by the polars of the cuboctahedron and the icosidodecahedron.

Figure 18 The complex $\mathcal{K}_m$ (m>4)

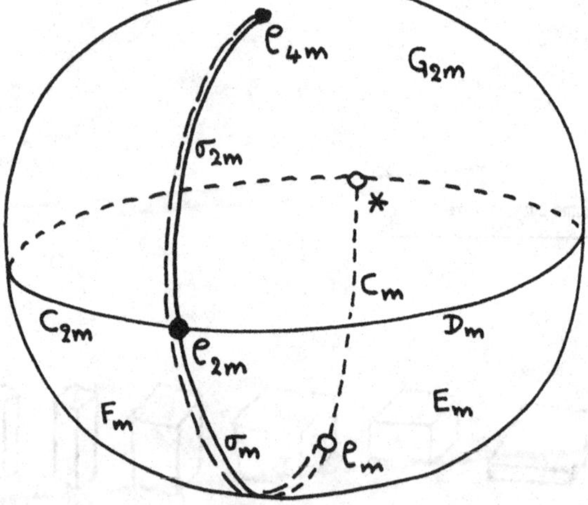

7. Edge-regular polyhedra

In §2.9 we referred to the fact that any combinatorially regular polyhedron is combinatorially equivalent to one of the five Platonic solids. We also showed that if a polyhedron is combinatorially edge-regular then it is combinatorially vertex-regular or it is combinatorially face-regular. Moreover, the orbit-vector of any edge-regular polyhedron P takes one of the forms (1,1,1), (1,1,2) or (2,1,1).

The first of these three possibilities is realised iff P is combinatorially regular. Suppose then that $\omega(P) = (1,1,2)$. Then the v vertices of P have a common valency 2a; each edge of P is incident with exactly one member of each face-orbit; and the face-orbits consist of h p-gons and k q-gons, where $h + k = f$. It follows immediately that either $p = 3$ or $q = 3$, since every polyhedron has either a triangular face or a vertex of valency 3 (see §2.9). For definiteness, say $p = 3$. Then

$$e = av = 3h = qk .$$

Combining these relations with Euler's relation, we find that

$$h\left(\frac{3}{a} + \frac{3}{q} - 2\right) = 2 .$$

Since $h \geq 4$, we conclude that

$$2 < \frac{3}{a} + \frac{3}{q} - 2 \leq \frac{5}{2} .$$

But $a \geq 2$ and $q \geq 3$, so the only solutions of these inequalities are

$$(a,q) = (2,3),\ (2,4) \text{ and } (2,5) .$$

Hence P is combinatorially equivalent to an octahedron, a cuboctahedron, or an icosidodecahedron, as shown in Figure 19. For the octahedron, $\omega(P) = (1,1,1)$, but we might imagine that the two families of faces are of different 'colours', say red and green, giving $\Diamond_3$ the courtesy status of the case $\omega(P) = (1,1,2)$.

To find those polyhedra P for which $\omega(P) = (2,1,1)$, we need only consider the polars of the cuboctahedron and icosidodecahedron, namely the rhombic dodecahedron of the first kind and the rhombic triacontahedron. The first of these is bounded by 12 identical rhombuses, in which the ratio of the longer to the shorter diagonal is $\sqrt{2}$. The second has 30 identical rhomboidal faces, in which this diagonal ratio is the golden number $\gamma = \frac{1}{2}(1+\sqrt{5})$. Of course the cube, as the polar of the octahedron, may be considered as a third instance, whose faces are rhombuses in which the diagonal ratio is 1 , and in which alternative vertices are given different colours or marks.

We have now identified the complete family of nine combinatorial types of combinatorially edge-regular polyhedra, each of which can be realised by a (metrically) edge-regular polyhedron. The relationships between these nine objects are extremely interesting, forming part of the theory arising from 'Wythoff's construction' [Coxeter, 1948].

Figure 19 The case $\omega(P) = (1,1,2)$

	v	e	f	p	q	h	k
'coloured' octahedron	6	12	88	3	3	4	4
cuboctahedron	12	24	14	3	4	8	6
icosidodecahedron	30	60	32	3	5	20	12

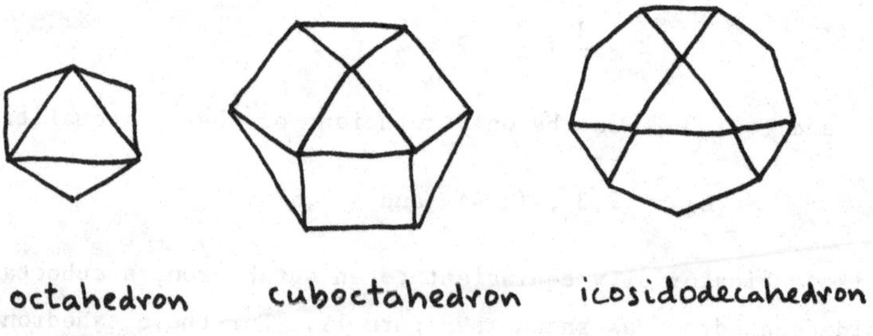

Let P be any of the Platonic solids, and let S be the sphere in E^3 with centre the centroid c of P and radius the distance ρ from c to the mid-point of any edge of P. The fact that P is Platonic implies that S touches each edge at its mid-point, and the polar $P_\rho^* = P'$ of P with respect to S is another regular solid having the same property with respect to S. If $\alpha : F(P) \to F(P')$ is the lattice anti-isomorphism given by polarity with respect to S, then each edge T of P intersects the edge $T' = \alpha(T)$ of P' orthogonally at their common mid-point. Thus T and T' are the diagonals of a rhombus T_+.

The polyhedron $P_\cap = P \cap P'$ has as vertices the mid-points of the edges of P (or P'), and $P_\cap$ is vertex-regular. The polar $P_+ = (P_\cap)_\rho^*$ of $P_\cap$ with respect to S is the polyhedron $\operatorname{conv}(P \cup P')$, whose faces are the rhombuses T_+, T ranging over the edges of P. Of course P_+ is face-regular.

Thus each of the three unordered pairs $[P, P']$, in which P, P' are mutually polar edge-regular polyhedra, yields an ordered pair $(P_\cap, P_+)$, where $P_\cap$ is vertex-regular and P_+ is face-regular. Thus $\omega(P_\cap) = (1,1,1)$ or $(1,1,2)$ and $\omega(P_+) = (1,1,1)$ or $(2,1,1)$. Table 20 displays the three cases, using the labels already employed in previous sections. (Note that in the first row, $P \cup P'$ is Kepler's 'stella octangula'.)

Table 20

	P	P'	P	P_+
A	tetrahedron	A tetrahedron	B octahedron	D cube
B	octahedron	D cube	C cuboctahedron	C* rhombic dodecahedron
E	icosahedron	G dodecahedron	F icosidodecahedron	F* triacontahedron

8. Perfect polyhedra

Consider the following propositions:

(1) P is perfect;

(2) P is edge-regular;

(3) $P = A,B,C,C^*,D,E,F,F^*$ or G ;

(4) Deicke's conjecture.

We have shown that:

(2) <=> (3) ;

(3) => (1) ;

(4) => {(1) <=> (2)} .

Thus (4) => {(1) <=> (3)} . Since no proof of (4) has been devised, we still need a proof that (1) => (3) . This can be done as follows. There is a finite list of polyhedra P , up to similarity, all of whose vertices are corners of a fundamental region of $\Gamma(P)$. Among these similarity classes, only the nine listed in (3) are nodes. But any polyhedron P having a vertex that is not a corner of a fundamental region of $\Gamma(P)$ has positive deficiency. Hence (1) => (3) .

Concluding Remarks

The preceding chapters contain only an outline of how a theory of symmetry for polytopes can be developed. It will be obvious to any reader that there remains a great deal of detailed work to be done. Perhaps the most obvious next step is to investigate 4-polytopes with respect to symmetry equivalence. Since all finite subgroups of O(4) are known (see, for example, du Val (1964)), it is in principle possible to describe the symmetry type structure of the families of vertex-transitive and facet-transitive 4-polytopes, and to determine all the perfect 4-polytopes. These tasks have not been carried out as yet, and it seems to me that it would be more worthwhile to search for general theorems. For instance, is there an analogue of Deicke's conjecture for 4-polytopes, and, more generally, for n-polytopes? We know already that there is a countable infinity of similarity classes of perfect polytopes, of the form $\ast(P \square P)$ and $\ast(P \diamond P)$, where P is any regular polygon, but there are certainly many others, besides the regular 4-polytopes, not of this form. The relation between such 4-polytopes and regular complex polygons (Coxeter, 1973) should also be investigated.

In dimension 5 (as in all higher odd dimensions), the outstanding question is whether the number of nodes (similarity classes of perfect 5-polytopes) is finite.

I should like very much to see some progress on the conjectures regarding polarity and duality. I have to admit, however, that I have at present little idea about how these might be tackled.

Although the polytope theory is so incomplete, it is difficult to avoid exploring the symmetry theory of more general objects. We might think of replacing the ambient Euclidean E^n space by, say, a space of constant curvature, or an arbitrary Riemannian manifold, or even just a metric space. We might even study symmetry in a general categorical setting (see, for example, Robertson (1976) or with respect to structures in the sense of Bourbaki (1968) A more modest programme is to look at objects more general than polytopes, retaining E^n as the ambient space. A first step in this direction is to consider convex bodies in E^n. I give here some indications of what might be done.

Let B be a compact subset of $E^n \subset E^\infty$ with affine hull $\operatorname{aff} B = X$ of dimension k, where $-1 \leqslant k \leqslant n$. We say that B is a <u>convex body</u> if it is both convex and connected, and we say that such a convex body B is a <u>convex</u> k-<u>body</u> or has <u>dimension</u> $\dim B = k$. The construction described in §1.6 can be used again here to endow the set $\mathcal{B}$ of all convex bodies with a suitable topology, in which the empty set forms a singleton component, while the set $\mathcal{B}' = \mathcal{B} \setminus \{\emptyset\}$ is a connected metric space. In taking similarity classes, it is again convenient to detach the convex 0-bodies (the same thing as 0-polytopes) as a separate, singleton component. Thus we obtain an arcwise-connected space $\mathcal{B}^+$ of convex bodies of positive dimension, and $\mathcal{B}^+/\sim$ is also arcwise-connected.

The partial ordering of $\mathcal{B}$ by inclusion may be refined to yield the concept of <u>face</u>, as follows. Let B and C be convex bodies. Then B is a <u>face</u> of C , written $B \lhd C$, iff $B \subset C$ and for all $x, y \in C$, $x \in B$ and $y \in B$ whenever $tx + (1-t)y \in B$ for some $0 < t < 1$. If $B \lhd C$ and $\dim B = k$, then B is a k-<u>face</u> of C . Every convex k-body B has a unique (-1)-face, namely   , and a unique k-face, namely B itself. However, the set $F_j(B)$ of j-faces of B may be empty for some $0 < j < k$. For example, $F_j(D^n)$ is empty for $0 < j < n$ for any $n \geq 2$, where D^n is the closed unit disc in E^n .

The set $F(B)$ of all faces of B is a lattice (not necessarily finite) graded by the dimension of the faces. Combinatorial equivalence $\approx$ is defined by putting $B \approx C$ iff there is a dimension-preserving isomorphism of lattices $\lambda : F(B) \to F(C)$. Among combinatorial invariants, the combinatorial automorphism group $\operatorname{Aut} B$ is of interest, but face-vectors, incidence matrices and multiplicity lose their usefulness with loss of finiteness.

Symmetry equivalence is defined exactly as for polytopes (but lattice isomorphisms must preserve the grading). We no longer expect, of course, that the symmetry type $\divideontimes B$ of a convex body B need have a manifold structure. Nevertheless, it makes sense to say that B is of <u>perfect form</u> (or is <u>perfect</u>) if $\divideontimes B$ is a singleton. What perfect convex bodies can we find?

Since the operations $\square$ and $\Diamond$ apply to convex bodies, and $\square^r B$ and $\Diamond^s B$ are perfect whenever B is perfect, we can manufacture infinitely many examples. Thus if $J^r_s = \square^r \Diamond^s$, where $r + s > 0$ and

$$B^{r_1 \ldots r_m}_{s_1 \ldots s_m} = J^{r_1}_{s_1} \ldots J^{r_m} B$$ is perfect for any perfect convex body B .

In particular, $(D^n)^{r_1 \ldots r_m}_{s_1 \ldots s_m}$ is perfect in this sense.

In the visual range of dimensions $(n \leqslant 3)$, this device yields only the cube $\square^3$, the octahedron $\diamond^3$ and the discs D^2 and D^3, which might have pleased Aristotle. The convex hull $D^2 \square D^2$ of the Clifford torus is perfect, as is the less familiar object $D^2 \diamond D^2$.

I mention two other examples of perfect convex bodies, which are of a rather different kind, and which give a hint of the rich variety to be expected. Firstly, put $E^6 = E^2 \times E^2 \times E^2$, and let B be the convex hull of the union of $D^2 \times D^2 \times 0$, $0 \times D^2 \times D^2$ and $D^2 \times 0 \times D^2$. Then B is perfect. (This is one of a whole family of constructions, the simplest case being that of the cuboctahedron.) This construction was pointed out to me by C.T.C. Wall. A second example, which I owe to H.R. Morton and I.R. Porteous, is the convex hull $\tilde{V}$ of the Veronese surface V in E^6. The latter consists of all points of the form $(x^2, y^2, z^2, xy, yz, zx)$ in E^6 that lie on the unit sphere. To prove that these, and other convex bodies, are perfect, it is enough to exhibit these objects as singleton orbit types in the orbit space of a suitable Lie group action. There is every indication, therefore, that perfect convex bodies are abundant but interesting.

One of the casualties of the generalisation from polytopes to convex bodies is the existence of polars or duals. Given a convex body B, it seems that in general there is no convex body B^* such that $F(B)$ is anti-isomorphic to $F(B^*)$, with grade-reversal.

Bibliography

Besides references to work cited in the text, this list includes a number of items that might be of interest to the reader of the present work. Several of the monographs listed contain extensive and valuable bibliographies.

Aleksandrov, A.D. [1950]. Convex Polyhedra (Russian). Moscow. (Konvexe Polyeder (German). Berlin, 1958.)

Biggs, N.L., Lloyd, E.K. and Wilson, R. [1976]. Graph Theory 1736-1936. Oxford: Clarendon Press.

Bilinski, S. [1960]. Über die Rhombenisoeder. Glasnik 15, 251-263.

Borel, A. [1960]. Seminar on Transformation Groups. Ann. of Math. Studies 46. Princeton: Princeton University Press.

Bourbaki, N. [1968]. Elements of Mathematics: Theory of Sets, Hermann, Paris; Addison-Wesley, Reading, Mass.

Bredon, G.E. [1972]. Compact Transformation Groups. New York: Academic Press.

Coxeter, H.S.M. [1948]. Regular Polytopes. New York: Macmillan. (3rd edition, New York: Dover, 1973.)

" [1973]. Regular Complex Polytopes. Cambridge: Cambridge University Press.

Cundy, H.M. & Rollett, A.P. [1951]. Mathematical Models. Oxford: Clarendon Press. (2nd edition, 1961.)

Fejes Toth, L. [1964]. Regular Figures. New York: Pergamon.

Grunbaum, B. [1967]. Convex Polytopes. New York: Interscience.

" , and Shephard, G.C. [1981]. Patterns on the 2-sphere. Mathematika, 28, 1-35.

" , and Shephard, G.C. [1981]. Tilings and Patterns. San Francisco: W.H. Freeman & Co.

Heath, T.L. [1956]. The Thirteen Volumes of Euclid's Elements. 2nd edition. New York: Dover.

Holden, A. [1971]. Shapes, Space and Symmetry. New York: Columbia University Press.

Jones, G.A. & Singerman, D. [1978]. Theory of maps on orientable surfaces. Proc. London Math. Soc. 37, 273-307.

Kelley, J.L. [1942]. Hyperspaces of a continuum. Trans. Amer. Math. Soc. 52, 22-36.

Kendall, D.G. [1983]. Shape manifolds, procrustean metrics and complex projective spaces, to appear in Ann. of Prob.

Lee, H.D.P. [1965]. Plato: Timaeus and Critias. London: Penguin.

Loeb, A.L. [1976]. Space Structures. Reading, Mass.: Addison-Wesley.

Lyusternik, L.A. [1956]. Convex Figures and Polyhedra (Russian). Moscow. (English translation, T.J. Smith, 1963. New York: Dover.)

McMullen, P. [1967]. Combinatorially regular polytopes. Mathematika 14, 142-150.

Montgomery, D. & Zippin, L. [1955]. Topological Transformation Groups, New York: Interscience.

Pont, J.-C. [1974]. La Topologie Algébrique des Origines à Poincaré. Paris: Presses Universitaires de France.

Pugh, A. [1976]. Polyhedra: A Visual Approach. Berkeley: University of California Press.

Robertson, S.A. [1976]. Mobility in categories and metric spaces. In Differential Geometry and Relativity: in honour of A. Lichnerowicz on his 60th birthday. eds. M. Cahen & M. Flato, pp. 147-158. Dordrecht: D. Reidel.

" [1977]. Classifying triangles and quadrilaterals. Math. Gaz. 61, 38-49.

" [1981]. Symmetry and perfection of form. Interdisciplinary Science Reviews, 6, 340-345.

" & Carter, S. [1970]. On the Platonic and Archimedean solids. J. London Math. Soc. 2, 125-132.

" , Carter, S. & Morton, H.R. [1970]. Finite orthogonal symmetry. Topology 9, 79-95.

Robinson, G. de B. [1931]. On the fundamental region of a group, and the family of configurations that arise therefrom. J. London Math. Soc. 6, 70-75.

Rose, J.S. [1978]. A course on Group Theory. Cambridge: Cambridge University Press.

Schwarzenberger, R.L.E. [1972]. Classification of crystal lattices, Proc. Camb. Phil. Soc. 72, 325-349.

" [1974a]. Crystallography in spaces of arbitrary dimension. Proc. Camb. Phil. Soc. 76, 23-32.

" [1974b]. The 17 plane symmetry groups. Math. Gazette 58, 123-131.

" [1980]. N-dimensional Crystallography, Research Notes in Mathematics 20, Pitman, San Francisco.

Shephard, G.C. & McMullen, P. [1971]. Convex Polytopes and the Upper Bound Conjecture. London Math. Soc. Lecture Notes 3. Cambridge: Cambridge University Press.

Schoute, P.H. [1896]. Het vierdimensionale prismoide. Verhand. der Kon. Akad. Wet. Amsterdam (1) 5.2, p. 20.

Stott, A.B. [1910]. Geometrical deduction of semiregular from regular polytopes and space fillings. Verhand. der Kon. Akad. Wet. Amsterdam (1) 11.1.

du Val, P. [1964]. Homographies, Quaternious and Rotations. Oxford: Clarendon Press.

Wenninger, M.J. [1971]. Polyhedron Models. Cambridge: Cambridge University Press.

" [1979]. Spherical Models. Cambridge: Cambridge University Press.

Weyl, H. [1952]. Symmetry. Princeton: Princeton University Press.

Wythoff, W.A. [1907]. A relation between the polytopes of the C_{600} - family. Kon. Akad. Wet. Amsterdam, Science section, 9, 529-534.

Zalgaller, V.A. [1966]. Convex Polyhedra with Regular Faces. (Russian) Leningrad: Steklov Institute. (English translation, 1969. New York: Consultants Bureau.)

Index of Symbols

Many ad hoc symbols, especially those used in Chapters 5 and 6, are omitted.

Index of Names

Names of authors to whom reference is made in the Bibliography but not in the text are omitted.

General Index

Printed in the United States
By Bookmasters